P9-DWU-777

Genetics of Original Sin

Published in association with Éditions Odile Jacob for the purpose of bringing new and innovative books to English-language readers. The goals of Éditions Odile Jacob are to improve our understanding of society, the discussions that shape it, and the scientific discoveries that alter its vision, and thus contribute to and enrich the current debate of ideas.

Genetics of Original Sin

The Impact of Natural Selection on the Future of Humanity

CHRISTIAN DE DUVE

WITH NEIL PATTERSON

FOREWORD BY
EDWARD O. WILSON

Yale
UNIVERSITY PRESS
New Haven & London

Éditions Odile Jacob
Paris

Published with assistance from the foundation established in memory of
Philip Hamilton McMillan of the Class of 1894, Yale College.

Translated from *Génétique du péché originel,* by Christian de Duve,
published by Éditions Odile Jacob in 2009. Copyright Odile Jacob, 2009;
ISBN 978-2-7381-2218-6.

Set in Minion type by Integrated Publishing Solutions.
Printed in the United States of America.

Library of Congress Cataloging-in-Publication Data

De Duve, Christian.
[Génétique du péché originel. English]
Genetics of original sin : the impact of natural selection on the future of humanity / Christian de Duve with Neil Patterson ; foreword by E. O. Wilson.
p. cm.
Includes index.
ISBN 978-0-300-16507-4 (clothbound : alk. paper) 1. Life—Origin. 2. Life (Biology) 3. Evolution (Biology) 4. Natural selection. 5. Genetics. 6. Twenty-first century—Forecasts. I. Patterson, Neil. II. Title.

QH325.D41313 2010
576.8—dc22
2010029161

A catalogue record for this book is available from the British Library.

This paper meets the requirements of ANSI/NISO Z39.48–1992 (Permanence of Paper).

10 9 8 7 6 5 4 3 2 1

ALSO BY CHRISTIAN DE DUVE

A Guided Tour of the Living Cell, Scientific American Books (New York: W. H. Freeman, 1984)

Blueprint for a Cell: The Nature and Origin of Life (Burlington, NC: Neil Patterson, Publishers, 1991)

Vital Dust: Life as a Cosmic Imperative (New York: Basic Books, 1995)

Life Evolving: Molecules, Mind, and Meaning (New York: Oxford University Press, 2002)

Singularities: Landmarks on the Pathways of Life (New York: Cambridge University Press, 2005)

To Janine

"And when the woman saw that the tree was good for food, and that it was pleasant to the eyes, and a tree to be desired to make one wise, she took of the fruit thereof, and did eat, and gave also unto her husband with her; and he did eat."

—GENESIS 3:6

Contents

Foreword

Christian de Duve has delivered a clear statement of why ours is the Century of Biology. If there is anything that science has taught us, it is that humanity is a biological species in a biological world. We originated here, grew up here, and are thoroughly adapted to this world in every fiber of our bodies and every neuronal circuit that thrums through our brain. In the fundamentals of structure and development, we are not different from other organisms. And in the fine details of anatomy, we are close to our phylogenetic cousins, the great apes.

With the smooth mastery acquired by a lifetime of distinguished scientific research, Professor de Duve guides the reader through 3.5 billion years of history that led from the earliest microbes to the present-day global biodiversity, including one of its most recent productions, the hominines—us. That said, let us not stress humanity's humble origins to the extent of devaluating the immense achievement they represent. Humans are not only the smartest creatures ever evolved, exceeding by a wide margin the nearest competitors (great apes, elephants, cetaceans), we are also the only species to create culture based upon, with each piece potentially immortal, an infinitely creative language. We alone are capable of endless histories, fantasies, and instructions.

We are nothing less than the mind of the biosphere. The achievement has been one of the major transitions of evolution, which together led from macromolecule to cell to eukaryotic cell to multicellular organism to society to the human-grade level of culture. However, whereas the earlier transitions occurred hundreds of millions or even billions of years ago, the last great transition, to the human level, occurred at best a few hundreds of thousands of years ago. It finally came to full flower in the Neolithic dawn a scant ten thousand years ago. Therein lies the dilemma identified by Professor de Duve. The earlier transitions occurred with agonizing slowness, while the human transition burst into the world as a biological supernova. Earth has not had time to adjust to this magnitude and abruptness—nor have we. The human condition is that depicted in the *Star Wars* movie trilogy: we have Paleolithic emotions upon which have been erected medieval institutions and godlike technology.

Having explained this dilemma in clear detail, de Duve then invites the reader to join in finding the solution or, better, ensemble of solutions. The fundamental premise in his exercise is that a knowledge of humanity's origins and nature, by scientists and the public alike, is necessary to find the correct solution. This is the transcendent goal, truly vital in nature, that requires the best that science, religion, and political leadership can put together.

Edward O. Wilson

Preface

Life is the most extraordinary and perfected natural manifestation known to us. It has not ceased, ever since human beings have existed, to inspire awe and wonder. And now, for the first time in the history of humankind, knowledge and understanding have been added to those sentiments. This is a new situation. A mere four hundred years ago, it was not realised that blood flows round in a closed circuit or that living beings are made of cells; no microbe had been seen. Two hundred years ago, it was not known that infectious diseases are caused by invisible forms of life; there were no vaccines (except against smallpox, empirically introduced in 1796); no antibiotics; and it was not yet appreciated that all living beings, from microbes to humans, are part of a large family tree, born from a single root more than three and a half billion years ago. As recently as sixty years ago, knowledge of the fine structure of cells, their chemical constituents, and the fundamental mechanisms that underlie their activities was still in its infancy. Virtually nothing was known about DNA. The terms "double helix" and "genetic code" had not been invented. Today, in the space of just my own lifetime, all of these vital facts and processes have been clarified. It is no exaggeration to say that we *understand life on Earth.* Many details remain to be elucidated, but the essentials are known. What is probably the greatest leap in the history of knowledge has been accomplished. Such an illumination must not be kept to a few initiates.

This is all the more true because it's not just about the life that surrounds us; it's about our own nature, our own history as a form of life. One of the most important revelations of modern science has been the discovery that we are part of the great network of life. We are not only its spectators and beneficiaries, as long believed. We are born from it and share its basic properties with all other living beings on this planet. In addition, we have specifically human traits that we owe to our brains. Understanding life means understanding ourselves.

There are more practical reasons why it is important for all of us, and in particular our political, cultural, economic, and religious leaders, to be informed about the nature and history of life. Our understanding of basic biological mechanisms has spawned powerful means to manipulate life. Cloning, in vitro fertilization, stem cells, DNA tests, genetically modified organisms: these terms and others have become part of common vocabulary and should be understood by everyone. This requirement does not just concern specialized notions. Many questions of interest for our daily existence—health, food, hygiene, economy, environment, and so on—are linked one way or another to what we know—or ought to know—about the properties of living beings. The term "bio" has gained an almost mystic connotation. But rare are those who may sensibly claim that they understand with some precision what it's all about.

The most important and urgent reason why it is now imperative that every responsible citizen be informed of recent developments in the life sciences is that *we need this knowledge* in order to face the future in a constructive way. The present book addresses this issue. It is the outcome of a thirty-year voyage of discovery, chronicled in five successive books, which has led me from lysosomes and peroxisomes, the cell organelles that were long the sole focus of my laboratory research, first to all aspects of cellular organization (*A Guided Tour of the Living Cell*), then to the basic properties of life and to its origin (*Blueprint for a Cell: The Nature and Origin of Life*), to its evolutionary history, including the advent of humankind (*Vital Dust: Life as a Cosmic Imperative*), and, finally, to the "meaning of it all" (*Life Evolving: Molecules, Mind, and Meaning*). At the end of the journey, I completed this series, which was largely addressed to the general public, with a concise summary aimed at a more scientifically literate readership (*Singularities: Landmarks on the Pathways of Life*). This was to be the end of my literary expedition. My neuronal telomeres decided differently. (Highlighted by the 2009 Nobel Prize in medicine, telomeres are DNA "tails" attached to the ends of chromosomes that shorten progressively in the course of successive cell divisions. Their preservation and/or repair are linked to increasing cellular longevity.)

Having been granted the time to do so, I have been prompted to go one step further, this time back to the past and into the future. It turned out to be not just a pleasing intellectual pastime but an effort to meet a pressing need. The future of life on Earth, in particular human life, is under serious threat. Climate change, deforestation, desertification, water shortage, famine, loss of biodiversity, depletion of natural resources, growing energy needs, pollution, new diseases, overcrowded megalopolises, conflicts, and wars are daily brought to our attention by the media. These menaces are recognized and burgeoning. Yet responses so far have been sluggish, to say the least. The vague and slight results of the recent Copenhagen conference on climate illustrate humanity's inability to come to grips with threats that extend beyond the immediate future.

It occurred to me that the causes of this worrisome situation, both the menaces and our lack of constructive response, are to be found in our nature, itself the product of a long history that I had attempted to reconstruct and explain in previous books. The "culprit," in this scientific interpretation of the myth of "original sin," is natural selection, which has sustained in our genes the traits that proved immediately useful to the survival and reproduction of our ancestors but have now become dangerously harmful. If we wish to escape the fate that awaits us, we must take advantage of our unique ability to consciously and deliberately act against natural selection. Such is the object of this book.

The work is addressed to as wide a readership as possible, not necessarily acquainted with science or trained in its way of thinking, but adequately informed of world affairs. For this reason, I have avoided all technical terms and left out the customary notes and references that would be needed in a more scholarly opus. I have also mostly refrained from mentioning quantitative data readily available elsewhere. But I do explain, in passing, some of the new biotechnological tools, such as cloning and genetic engineering, offered by modern science to a society that distrusts them and urgently needs clarification on the topic.

I cannot end this preface without recalling with deep sorrow the memory of my dear Janine, lost to my affection after sixty-five years of life together, just as I was finishing the first draft of this book. She used to read each chapter as soon as it came out of my computer and never failed to make judicious comments, all the more valuable because, as a professional artist, she cast a fresh look on my writings. I dedicate what is probably my last work to her beloved memory, on the day of what would have been her eighty-eighth birthday.

Nethen and New York, 6 April 2010

Acknowledgments

As I have done for earlier works, I wrote this book concurrently in my two mother languages, English and French. The two versions have benefited reciprocally from the criticisms, comments, and suggestions addressed to the other.

Many people have helped to make this work possible. Among them, I owe special tribute to my faithful editor and longtime friend Neil Patterson, who once again has favored me with his invaluable help, not only in purging my style of gallicisms, grammatical errors, obscure statements, needless repetitions, and flowery expressions, but also in critically assessing the substance. His participation has far exceeded a strictly editorial assistance and has justified his designation as coauthor. I am particularly pleased to publish this new work with him, after more than twenty-five years of friendly and effective, if sometimes heated, collaboration.

I am particularly indebted to my valued friend Odile Jacob, who not only has published two of my previous books before this one but has, in addition, decided to include my latest opus in her joint enterprise with Yale University Press. It is a great honor. I also address my most grateful thanks to the members of her staff, including Gérard Jorland, Émilie Barian, and Claudine Roth-Isler, for their excellent collaboration.

I owe a similar debt of sincere gratitude to Yale University Press for publishing the English version of the book and, especially, to Jean Thomson Black for her competent and understanding assistance, to Laura Jones Dooley for her thorough editorial revision of the manuscript (with my apologies for not always following her recommendations), and to Jaya Chatterjee for her valuable help with illustrations.

Finally, I wish to thank my trusted Brussels assistant, Monique Van de Maele, for her invaluable help in the search for needed information and her colleague, Nathalie Chevalier, who has expertly assembled the illustrations common to both versions. I am also indebted to Xavier de Felipe for his wonderful reconstruction of the "forest of neurons" reproduced in figure 10.1 and to Gabriel Ringlet for valuable criticisms and suggestions.

Introduction

The sacred writers who invented the famous myth, immortalized by numerous artists and writers across the centuries, of Original Sin that allegedly cost the first parents of humanity to be expelled from the Earthly Paradise, have not just displayed lively poetic imagination. They have, in addition, shown remarkable perspicacity—apart from their choice, which was far from innocent, of a woman as culprit. They have perceived the presence in human nature of a fatal flaw, which, as they saw it, only divine intervention could possibly repair. Hence the hope for an envoy from God, a Messiah, Savior, or Redeemer, whom some believe to have appeared two thousand years ago and others are still awaiting.

Modern science has rendered the biblical account untenable, without, however, invalidating the intuition that may have inspired it. Humankind is indeed tainted by a fundamental defect, bound, in all probability, to bring about its demise. The culprit is not Eve, but natural selection. There is indeed a need for a redeemer. But that redeemer will not come from heaven; only humankind can serve that role.

In this predicament, the wisdom of our forebears is of little help to us today, because the wise of yesteryear could not foresee the present crisis. But their recommendation that we should take advantage of the lessons of the past to prepare the future remains timely. What humanity needs now is a new form of wisdom inspired by what we have learned about the nature and history of the living world to which we belong, about our place in it, and about the manner in which we have reached it.

Such is the thesis I develop in this book, adopting as main guide life itself, as we have learned to understand it, with as illuminating beacon natural selection, the mechanism discerned by the genius of Charles Darwin, who has been celebrated in 2009 by a dual anniversary, the 200th of his birth and the 150th of the first publication of his magnum opus, *On the Origin of Species by Means of Natural Selection, or the Preservation of Favoured Races in the Struggle for Life.*

This book is divided into four parts, each of which stands more or less on its own. To start, I retrace briefly a number of basic notions about the nature, origin, and evolutionary history of life on Earth. I do so in a strictly descriptive fashion, leaving an analysis of the underlying mechanisms to the second part.

In Part II, I discuss such key processes as metabolism, reproduction, and development before addressing the central theme of the book: natural selection. I end with a brief consideration of some of the evolutionary mechanisms that have been proposed besides natural selection.

The scene thus set, I move on, in Part III, to the extraordinary saga of the human adventure, which, initiated a few million years ago in the heart of Africa, has developed, sustained by a stupendous expansion of the brain, at an increasingly dizzying pace, leading, in the last centuries and, even more so, in the very last decades, to the fantastic success of our species and to the mortal menaces it causes to weigh on the future, the ultimate consequences of our "original sin."

Then, in Part IV, I sketch potential solutions toward redemption or, at least, the chance for it, which I see in the specifically human power to act against natural selection. But to exercise this power, we will have to find in the resources of our minds a wisdom that is not written in our genes.

I
The History of Life on Earth

Introduction

What is life? What are its principal properties? What reasons do we have for believing there is only one kind of life on Earth, issued from a single ancestral root? How did life arise? What are the main steps of its history? Such are some of the questions that I try to answer in this first part of the book. I do so in descriptive and phenomenological form, postponing until the following part an examination of underlying mechanisms.

1
The Unity of Life

All known living organisms are descendants from a *common ancestral form of life,* often represented by the acronym LUCA (Last Universal Common Ancestor). Put forward as an affirmation, not just a theory or hypothesis, this statement may strike many readers unacquainted with modern biology as almost incredible, if not objectionable or even contrary to their most sacred beliefs. An explanation is in order.

Advancing knowledge has swept away "centrisms"

For most of their history, humans have seen Earth as the center of the universe, their privileged abode. In the second century, the Greek mathematician and astronomer Ptolemy integrated this "geocentric" view into a coherent theory that placed the Sun, the planets, and the stars in concentric spheres surrounding Earth. The Ptolemaic system dominated thinking for about fourteen hundred years, until the Polish astronomer Copernicus (1473–1543) rejected it in favor of the "heliocentric" view, which has the Sun in the center and the Earth and other planets circling around it. This view ran the risk of being seen as heretical at the time; it conflicted with the biblical account that Joshua "stopped the sun" to allow the Israelites to win the Battle of Gibeon against the Canaanite kings. For this reason, Copernicus prudently refrained from having his theory publicized until after his death. This may have been a sound decision, considering the fate that befell the Italian Galileo (1564–1642), almost one century later, when his advocacy of the Copernican view led, in 1633, to his condemnation by the Catholic Church, which, although yielding to the evidence much earlier, officially revoked this condemnation only some three and a half centuries later.

Since the time of Galileo, the Sun itself has lost its central status. It has been found to be only one among some one hundred billion stars in our galaxy, which has itself been dethroned by the observations of the American astronomer Edwin Hubble, who discovered, in the 1920s, that the distant celestial objects then known as "nebulae" are actually other galaxies, of which about one hundred billion are believed to exist.

While the status of our planet was progressively relegated from the center of the universe to the backyard of one in one hundred billion stars in one of one hundred billion galaxies, the "anthropocentric" view of a universe made for humans was also shaken. It all began with increasing realization that Earth is not, as was long believed, a fixed setting created for our human adventure. Earth was found to have a history of its own.

Earth has a history

This awareness dawned in the eighteenth century, from observations in places where flowing water has cut through rocks to expose their structure—the Grand Canyon is the most spectacular example—showing that the ground beneath us is stratified in layers of different texture and composition. The layers may be flat or curved, or inclined. Some contain the shells of marine animals embedded in the rock; think of marble, for instance. This telling clue enforced the astonishing conclusion that these layers had once been under water, where they had slowly formed through sedimentation of sand and dust particles, in which dead animals became buried, their bodies rotting, leaving only the mineral shells. As time went by and new layers accumulated on top, the old ones were pushed deeper and deeper, exposed to increasing heat and pressure, solidifying into rocks. Some of these rock layers were later driven upward by underground movements, to finally rise above the level of the waters in which they were born, even building mountains.

These observations allowed terrains to be classified in terms of their relative age. These ages could not be known in absolute terms at that time, but it was clear that they must cover considerable durations, possibly measured in as much as millions of years, a span of time almost unimaginable in those days. But the facts were there. Mountains obviously do not arise overnight, not even in a few millennia. The Alps have probably not changed much since Hannibal's army crossed them on elephants, more than two millennia ago, to take the Romans by surprise. Today, thanks to a method known as radioisotopic dating, geological times have been measured and found to be even longer than was first pictured, covering up to several billion years.

Although rudimentary in comparison with present-day geological knowledge, these early findings were sufficient to throw a revealing light on another set of observations that had long intrigued nature watchers, those dealing with fossils. Many such vestiges had been discovered by amateur naturalists over the years in various places. Their origin was a matter of lively debate. They were readily identified as the remnants of dead plants and animals, many of which, however, seemed to be different from any extant species known. Whether the living precursors of the fossils were organisms that still lived elsewhere in unexplored areas or had become extinct was much discussed. Some even considered that fossils were the remains of victims of the Flood. Putting some order into all this fantasizing, a crucial piece of information was provided by geological observations revealing that the complexity of the organisms whose fossils were found in a given terrain was related to the age of that terrain. The younger the terrain, the more complex the fossilized remains. These findings showed that life, like the Earth, also has a history, in the course of which organisms of increasing complexity progressively appeared.

Life also has a history

The notion of the "fixity of species" was, however, so dominant at the time that the significance of these observations was not immediately appreciated by most scientists. Only a few were sufficiently perceptive, as well as daring, to conceive the revolutionary hypothesis that life started with very simple forms that progressively evolved into forms of increasing complexity. This so-called transformist hypothesis was first formulated at the end of the eighteenth century, simultaneously in France, by Jean-Baptiste de Monet, chevalier de Lamarck (1744–1829), and, in England, by Erasmus Darwin (1731–1802), the grandfather of the famous Charles.

Highly controversial at the time it was first proposed, this theory has since been abundantly confirmed and is now established fact. Behind the extraordinary diversity of life-forms that make up what is known as the "biosphere," there lies an impressive set of similarities that all point to a single origin.

All living beings share a number of basic properties

All living organisms, from the simplest bacteria to humans, consist of one or more cells, which are microscopic entities enclosed by a membranous envelope and endowed with the ability to subsist under appropriate conditions, to grow, and to multiply by division.

All cells are constructed with the same molecular building blocks—mostly sugars, fatty acids, amino acids, nitrogenous bases, and a few mineral components—which are themselves assembled into the same kinds of large molecules, including polysaccharides (carbohydrates), lipids (fats), proteins, and nucleic acids (DNA and RNA).

All cells manufacture these constituents by the same chemical mechanisms—the bacteria in our gut and the nerve cells of our brain make their proteins in the same manner. All cells depend on the same types of metabolic reactions and use similar mechanisms to extract energy from the environment and convert it into work. There are differences, of course—plants derive their energy from sunlight, animals from the combustion of foodstuffs—but very quickly the two mechanisms converge into a common pathway (see chapter 4).

Even more impressive, all cells use the same genetic language. They all use DNA as repository of their genetic information, replicate this DNA by the same mechanism whenever they prepare to divide, and execute the instructions stored in the DNA by the same processes.

DNA (deoxyribonucleic acid) molecules consist of long chains made of a very large number of small molecular units, called bases, of which there are four different kinds, represented by their initials: A, for adenine, G, for guanine, C, for cytosine, and T, for thymine. The order, or sequence, in which the bases follow each other specifies the molecule's information content, just as the sequence of letters in a word specifies the word's information content. Our words are short but can carry large amounts of information because they are constructed with an alphabet of twenty-six letters. The DNA "alphabet" has only four "letters," but DNA "words" are very much longer than ours, often consisting of thousands of "letters." Their information capacity greatly exceeds that of our vocabularies.

The sole function of DNA is the storage of genetic information in a form capable of being *copied* by a mechanism, called "replication," which takes place every time a cell prepares to divide into two daughter cells, each of which will contain one of the two DNA copies. This phenomenon ensures the hereditary transmission of genetic information.

For this information to be turned into action, it must be transferred to RNA (ribonucleic acid), a closely related molecule, likewise constructed with an "alphabet" of four "letters": A, G, C (as in DNA), and U (for uracil, a substance very like T). The synthesis of RNA molecules on DNA templates is appropriately called "transcription" (the two languages are very similar).

The RNA molecules arising in this way carry out several functions. Many act as "messengers" instructing the synthesis of proteins, which, through their structural and catalytic properties, are the main agents that carry out the instructions transcribed from DNA to RNA. This function was first considered the principal, if not the only biological role of RNA. Later, however, it was discovered that some RNA molecules exert a catalytic function in some important processes, including the synthesis of proteins. Even more recently, it has been found that much of the DNA that does not code for messenger or catalytic RNAs, rather than being "junk," as was believed, actually codes for a large number of small RNA molecules endowed with a variety of regulatory functions. This has become one of the most fecund fields of research.

Proteins are also long molecular strings but made with twenty different kinds of units, called amino acids. They are molecular "words" made with an "alphabet" of twenty "letters." In the synthesis of proteins, known as "translation" (the two languages are totally different), the sequence of bases in the messenger RNA, which itself reflects the sequence of bases in the corresponding DNA, dictates the sequence of amino acids in the synthesized protein, according to a "dictionary," called the "genetic code." With minor exceptions due to late changes, this code is the same throughout the living world. Life is truly one; all forms of life are related.

The history of life is written into molecular sequences

To top it all, for those still not convinced by all those proofs, there is the incontrovertible evidence provided by the comparative study of the sequences of DNA genes, or of their RNA transcripts, or of their protein translation products. We have just seen that the information held by these molecular "words" is determined by the order, or sequence, of their molecular "letters," their spelling, so to speak. The last few decades have witnessed the development of techniques of extraordinary efficiency for deciphering those sequences, to the point that many entire genomes have now been sequenced, including the human genome, which contains some three billion "letters," the equivalent of about 150 volumes of the *Oxford Concise Dictionary!* This technology has revealed that genes that carry out the same function in different organisms show many sequence similarities, many more than could be accounted for by chance. The genes are unmistakably related and are all derived from a single ancestral gene by a pathway that has involved a number of changes in sequence (mutations), somewhat like words whose spelling has changed over time.

Not only have the sequence similarities been illuminating, by showing the single ancestry of many genes. The sequence differences have also been revealing. They have allowed the reconstruction of what is known as the "phylogenetic" (from the Greek *phylon*, race) history of the genes, that is, what amounts to their "genealogical tree," by a technique that uses the number of sequence differences between two forms of the same gene (belonging to two different organisms) as a measure of the time that has elapsed since the two genes separated from their last common ancestor and started evolving separately. Etymological research follows a similar line.

This method has been applied to a very large number of genes and continues to be applied more and more. Its results have confirmed—and sometimes corrected—a number of the conclusions derived from the study of fossils; especially, they have enormously enriched those conclusions. Indeed, the beauty of comparative sequencing is that it can throw light on the evolutionary history of any organism, not only those that have left fossil remnants. Fossils remain invaluable clues, of course, as illustrated by a number recently unearthed in China that have revealed several "missing links." But the innumerable organisms, such as soft-bodied animals and, especially, bacteria and other unicellular organisms, that have disappeared without trace can be reconstituted by the magic of molecular sequencing.

The history of life is written in the genes of extant organisms. It is written in very fine print, which it has been our generation's privilege to discover how to decipher. The general conclusion of all that has been learned is clear and indisputable: all known living organisms are descendants from a *single common ancestral form.*

Biological evolution is an established fact

The notoriously cautious language of science is rarely so affirmative. But, in the present case, with all the debates that surround the so-called theory of evolution, it is necessary to speak out unambiguously. Evolution is not a theory, contrary to what is often stated, sometimes even by scientists. *Evolution is a fact.* It was a theory two centuries ago, when Lamarck and Erasmus Darwin first proposed it, just as heliocentrism was a theory in the days of Copernicus and Galileo. Evolution is no longer a theory, just as heliocentrism is no longer a theory; it is a fact. The Catholic Church has recognized this with remarkable promptness, as compared to the Galileo affair. On October 22, 1996, at a meeting of the Pontifical Academy of Sciences, Pope John Paul II solemnly announced that "*evolution is more than an hypothesis.*" He did admittedly retreat somewhat to make a special case for the creation of the human soul; and his successor has retreated even further by leaning in favor of the so-called theory of intelligent design (see chapter 8). Nevertheless, biological evolution is not negated by the Catholic Church. Such is not the case for several other religious groups.

Opposition to evolution on religious grounds is widespread

Ever since Darwin, the notion of evolution has provoked opposition from religious groups. At one end of the spectrum are a number of fundamentalist Protestant Churches, especially in the United States, that deny evolution because it conflicts with what is written in the Bible, held to be directly inspired by God and, therefore, literally true. In line with this belief and in spite of all the evidence to the contrary, they persist in affirming that the world was created by God in seven days, some five thousand years ago, as written in Genesis. They will not recognize that the Bible, at least the early parts of it, was written almost three thousand years ago by human beings, possibly inspired by God if that is the cherished belief, but using the knowledge and language of their time.

At the other end of the spectrum are the defenders of so-called intelligent design, who pretend to invoke no explicit religious motivation but merely claim that purely natural phenomena cannot account for all evolutionary events. There will be more about this "creationism in disguise" in chapter 8.

Between these two extremes, there is a form of creationism—sometimes referred to as "old Earth" creationism, as opposed to the "young Earth" variety based on a literal interpretation of the Bible—that similarly denies evolution and advocates instant creation of living species, but on a more flexible time scale, consistent with the fossil record. This form of creationism is more widespread and professed by members of other Christian churches, including some conservative Catholics, notably in Poland, and also by many Muslims and by a number of Orthodox Jewish scholars.*

Defenders of this form of creationism often accept so-called micro-evolution, which takes place within existing genera, but deny "macro-evolution," the kind whereby new species arise from old ones. They use a number of allegedly scientific arguments to affirm, for example, that there is no valid proof of a descent of birds from reptiles or of reptiles from fish. As in

* See J. Sechbach and R. Gordon, eds., *Divine Action and Natural Selection: Science, Faith and Evolution* (Singapore, 2009).

the days of Darwin, the origin of humankind from chimpanzee-like ancestors is the most contested aspect of modern evolutionism. This is where the official voice of the Catholic Church departs from the scientific account of evolution. The biological descent of humans is accepted; but creation of the human soul is seen as a special event.

Creationism is not just a religious creed. It claims to be a science, which deserves to be taught alongside evolutionary biology, or even in place of it. Its proponents have built powerful organizations in pursuit of this goal. They fight in court and try to convince legislatures to give equal weight to what they call "creation science" in school curricula, or even to abolish the teaching of evolutionary biology. They provide teachers, not only in the United States, but also in other countries such as Poland or Turkey, with costly, beautifully illustrated "textbooks" in which the facts of life are reinterpreted within a creationist framework. They also try to propagate their ideas among the general public by, for example, filling the bookshops around the Grand Canyon with pseudo-scientific pamphlets describing this beautiful illustration of the Earth's history as a recent product of the Flood.

Further discussion of this phenomenon, which is the object of numerous books and debates, does not belong in this book. Let it simply be said that it conveys the image of a Creator who deliberately filled the world with all sorts of false clues that lead scientists astray, including the geological strata, the fossils, the kindred DNA molecules with which phylogenetic trees are constructed, the radioactive isotopes that allow us to date the Earth's history, and all the other pieces of evidence that rigorous and honest investigators have collected and used in their efforts to understand nature. Such an image of the Deity as a willful mystifier is hardly one a sincere believer is likely to defend.

2
The Origin of Life

The beginnings of life on Earth are shrouded in the darkness of a very distant past, going back at least 3.55 billion years—more than three and a half million millennia!—according to microscopic traces believed to be of fossilized bacteria, detected in rocks of that age. It is interesting to place this event within the framework of the history of our planet and of the history of the universe.

Life appeared on Earth shortly after the young planet had become physically able to harbor it

The Big Bang, the primeval explosion taken by most cosmologists to have sparked our universe into being, took place 13.7 billion years ago according to the most recent estimate. The solar system was born some 4.55 billion years ago—when the universe was already more than 9 billion years old—from a swirling cloud of gas and dust that gradually condensed into the central Sun and surrounding planets, including the Earth. This birth was a violent affair, which subsided only about 4 billion years ago, when the Earth became covered with bodies of liquid water and became, for the first time, physically capable of harboring life. Less than half a billion years later, maybe much earlier but leaving no record so far discovered, life was there. It is not impossible that life appeared as soon as the Earth was physically ready to receive it, or almost.

The origin of life is not known, but the only scientifically acceptable hypothesis is that it arose naturally

How life started is the object of much research and even more speculation. Instant divine creation is one possibility, not only advocated by creationists but also implicitly accepted by a large number of laypeople, perhaps a majority, who see life as due to some kind of "vital spirit" that was initially "blown" into matter and still goes on "animating" it in every living being. Everyday language is permeated with this belief.

Unlike strict creationism, this view, known as "vitalism," is not incompatible with evolution; it dominated biology for a long time, especially in France, where it was defended by many famous scientists, including Lamarck, one of the fathers of evolutionism, the celebrated Louis Pasteur, and, more recently, many other biologists, influenced by the philosopher Henri Bergson, winner of the 1927 Nobel Prize for literature, whose major opus, *L'Évolution créatrice,* recognized evolution, as the title says, but saw it as the product of an "élan vital," a vital surge. Remarkably, the one-hundredth anniversary of the publication of this book was celebrated in France with some prominence in 2007, in spite of its outdated character. Today, vitalism is rejected by most scientists, with the exception of the advocates of intelligent design, who espouse the related theory of finalism (see chapter 8). Thanks to the revolutionary advances of the last fifty years, we now understand and explain life entirely in natural terms.

The same can't be said of the origin of life, which is unknown so far. It thus remains permissible, while rejecting vitalism, to imagine, as some do, that life was flipped into being by a Creator, who subsequently left it to function and evolve under its own power, although such a conception of the deity does not fit with the more usual one of an omnipotent God who, notably, can be asked to change the course of things. As long as the origin of life can't be explained in natural terms, the hypothesis of an instant divine creation of life cannot objectively be ruled out. But this hypothesis is sterile, stifling any attempt to investigate the origin of life on Earth by scientific means. The only scientifically useful hypothesis is to assume that things, including the origin of life, can be naturally explained. If we start with the premise that they cannot, we may as well close our laboratories. Searching for an explanation that is taken, a priori, not to exist is futile (see also chapter 8).

The building blocks of life arise spontaneously throughout the universe

So far, investigations based on this "naturalistic postulate" have failed to provide an answer to the problem of the origin of life but have achieved some progress. One of the most important findings of the last decades is that the small molecular building blocks of life, the sugars, amino acids, fatty acids, and nitrogenous bases from which are constructed the larger polysaccharides, lipids, proteins, and nucleic acids that make up the bulk of so-called living matter, arise spontaneously in various sites of our solar system and, probably, in many other parts of our galaxy, as well as in other galaxies. These astounding facts, which belie the traditional view of *organic* chemistry as the prerogative of living *organisms,* were recently revealed by the spectral analysis of the radiation coming from outer space, by the probing of comets with instruments borne by spacecraft, and, especially, by the analysis, with all the resources of modern chemistry, of meteorites that have fallen on Earth.

Thus, what most likely constitutes the first stage in the origin of life is known. It is provided by cosmic chemistry, which, in innumerable parts of the universe, spontaneously generates the basic building blocks of life. Note that cosmic chemistry is not bioselective. It makes a gamut of organic compounds, of which some happen to participate in the building of living organisms, whereas many others do not. Subsequent events in the development of life have entailed a selection among the potential building blocks provided by cosmic chemistry.

Earth formed a "cauldron" in which cosmic building blocks could interact

From this naturalistic perspective, the products of cosmic chemistry are seen as landing in a milieu where some started interacting with each other to produce molecules of increasing size and complexity, which then interacted to produce polymolecular assemblages of increasing size and complexity, up to forming entities that could be defined as "protocells," or primitive cells. Here is the snag. Nobody has so far succeeded in reproducing such a situation, or even a small part of it, in the laboratory.

One problem is that there is no agreement yet on what may have been the milieu in which it all started. Some believe sunlight may have been needed as a source of energy. Others think that life originated in the darkness of deep waters. Debates also occur between proponents of a "hot cradle" and proponents of a "cold cradle." According to the latter, the constituents of life are much too fragile to be able to survive long enough in a hot environment. The former, on the other hand, have been influenced by the fact that all the bacteria identified as most ancient by molecular phylogenies are thermophilic, that is, adapted to very hot environments.

Another point of disagreement concerns the relationship between the early chemistry that led to life and present-day biochemistry. Many specialists, impressed with the total reliance of biochemistry on the activity of protein catalysts, or enzymes, which obviously are too complex to have been present at the dawn of life, argue that there is no relationship between the two chemistries. Some, however, including myself (see chapter 4), see biochemistry as flowing congruently from that early chemistry.

As to the temperature that surrounded life's birth, I adopt the opinion, held by a number of experts in the field, that life probably started in hot volcanic waters, perhaps in one of those underwater formations, called deep-sea hydrothermal vents, or black smokers, that spew overheated, pressurized, sulfurous, metal-laden waters from fissures in the bottom of oceans and have been found, against all expectations, to harbor many strange forms of life.

My reason for subscribing to this view does not, however, rest on the thermophily of the most ancient bacteria, which are probably late products of a long evolutionary process and may not tell much about the conditions under which life first arose. I see volcanic waters as the likely site where life originated because present-day processes of biological energy transfer universally use derivatives of two central compounds, inorganic pyrophosphate and hydrogen sulfide, that are produced naturally only in a volcanic environment.

The first steps in the origin of life were chemical in nature

Whatever the pathways involved in those early stages, they must have been *chemical* in nature, which means that they were bound to happen under the physical-chemical conditions that prevailed where they took place. Chemistry deals with strictly deterministic, reproducible phenomena. Were it not so and should chemical processes involve even a tiny element of chance, there would be no chemical laboratories, no chemical factories. We could not afford the risk. Thus, if early events leading to the origin of life were chemical, as they must have been if our premise of a natural—and not supernatural—origin of life is correct, then they were bound to occur under the prevailing conditions. It follows that if the same conditions should occur elsewhere in the universe, one would sensibly expect life to emerge similarly there, an implication of interest with respect to a related question much in vogue: Does extraterrestrial life exist?

This conclusion holds for all the early events in the origin of life, up to the appearance of the first information-bearing molecule capable of being replicated, that is, of inducing the making of copies of itself by whatever chemical machinery is responsible for making such molecules. As we shall see in chapter 7, this ability automatically entails the occurrence of selection, adding a new, chancy element to chemical determinism.

The appearance of RNA was a key step in the origin of life

In present-day life, the function of storing information in replicable form is universally fulfilled by DNA. There is, however, strong reason to believe that, in the origin of life, DNA has been preceded in this function by RNA, which also preceded proteins. To look into the arguments that support this opinion would take us into too many details. I mention it only in order to emphasize the crucial importance of the appearance of RNA in the origin of life, a veritable watershed separating a first stage exclusively dominated by chemistry from a second stage in which selection was added to chemistry (fig. 2.1). Experts in the field recognized this and have, in the last forty years, expended enormous efforts to try reproducing RNA synthesis under plausible early-Earth conditions. Many interesting results have been recorded; but the problem remains unsolved. Nobody has yet succeeded in making RNA in the laboratory, even under much more advantageous conditions than probably existed at the beginning of life. It could be said that a solution to this problem represents the "Holy Grail" of research on the origin of life.

Making RNA would be only one of a long series of steps. Another, of fundamental importance, would be the birth of proteins, most likely through the operation of RNA molecules, which are universally responsible for the synthesis of proteins in present-day living beings. Proteins would inaugurate enzymes and, with them, metabolism. I shall return to these questions in greater detail in chapter 4, devoted to metabolism. In addition, the function of storing genetic information in replicable form would have to be transferred from RNA, with which it first arose, to DNA, which accomplishes it today.

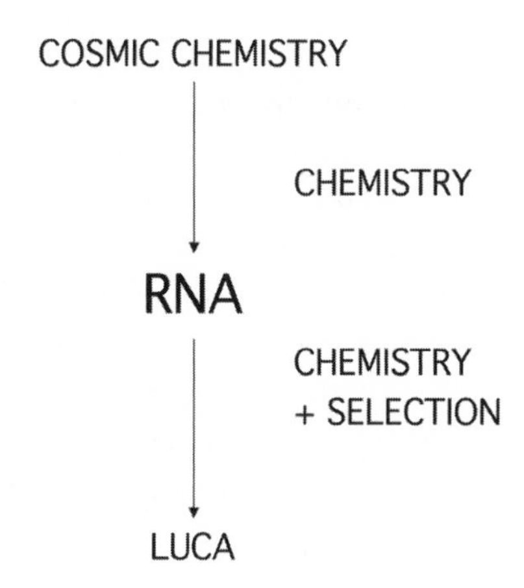

Fig. 2.1. *The origin of life.* This process may be defined as the chemical pathway, so far unknown, that has led from certain organic products, which form everywhere in the universe, to the last universal common ancestor (LUCA) of all life on Earth. This pathway may be divided into two stages separated by the appearance of RNA (or, in a more general fashion, of the first replicable information-bearing molecule). The first stage must have depended exclusively on chemistry. In the second, selection was added to chemistry (see chapter 7).

Finally, at some undetermined stage, these systems would have to become enclosed within an envelope, or membrane, to give rise to the first protocells.

All these events must have taken place when life first arose on our young planet. How and in what order is totally unknown. Also unknown is the manner in which the first protocells progressively evolved into the last universal common ancestor of all living beings, or LUCA, which, must, by definition, have possessed all the main properties living organisms have in common and no doubt inherited from this common ancestor.

There are plenty of challenges for future investigations. Whether these challenges will ever be successfully overcome cannot be predicted at this time. Today, the prospects seem bleak. On the other hand, experience has shown that a single breakthrough sometimes suddenly opens immense fields to scientific exploration. This has happened time and again, often with people exclaiming a posteriori: "Why didn't I think of that?"

3
The Evolution of Life

Born more than three and a half billion years ago, life remained unicellular during more than two and a half billion years, more than two-thirds of its existence on Earth. During all that time, microbes, organisms visible only with a microscope, were the only ones present on our planet. They are still abundant, occupying a wide variety of sites and revealing their presence by many effects, such as producing diverse chemical substances, causing many infectious diseases, and putrefying dead organisms (a function they share with certain molds).

Microbes have left few fossil vestiges but many other traces of their long duration on Earth

Very few fossil remains landmark the history of microbes on Earth, but many other signs of their passage have been left. They remodeled their habitat, leaving many traces found in geological structures and in the atmosphere today. And they have undergone a number of changes that have become known to us through the study of the genomes of extant organisms. Three major events highlight this long history.

Bacteria separated into two main groups

First, the initial lineage soon split into two major groups, known today as Archaea and Bacteria or, more commonly, as archaebacteria and eubacteria, which, together, make up what much of the world still calls "bacteria" and experts designate as "prokaryotes," organisms that do not possess a true nucleus (*karyon* in Greek).

Archaebacteria, so named because they were believed, perhaps wrongly, to be the more ancient of the two groups, include, among many others, two particularly fascinating classes of extant microorganisms, the methanogens and the extremophiles.

Methanogens are present wherever oxygen is absent and hydrogen is produced. There they survive by converting carbon dioxide (CO_2) and hydrogen (H_2) to methane gas (CH_4) and water (H_2O), a reaction that supplies them with enough energy to build their substance entirely from simple mineral building blocks. The mud at the bottom of ponds is one of their favorite habitats. The methane they produce in those murky depths creates the bubbles that break the silence of swamps by their muffled plopping; it fuels the will-o'-the-wisps that flit on the surface of marshes at night. Methanogens also thrive in the digestive tract of cattle and other ruminants. Methane is a greenhouse gas, which joins with carbon dioxide in the formation of the atmospheric shield that prevents heat from escaping and thereby contributes significantly to the warming of the global climate.

Extremophiles, the other remarkable class of archaebacteria, include organisms that are, as their name indicates, adapted to extreme physical conditions: elevated temperature, up to more than 110°C (230°F), icy cold, high pressure, burning acidity, caustic alkalinity, concentrated brine, not counting the innumerable human-made pollutants. Extremophiles illustrate, more than any other living beings, the amazing ability of life to respond to environmental challenges.

Eubacteria, the second group of prokaryotes, include most of the pathogenic varieties that cause such infectious diseases as tuberculosis, diphtheria, plague, leprosy, meningitis, pneumonia, and many others. They also comprise a large number of harmless varieties found in many natural niches, including the human gut, whose main inhabitant, *E. coli* (short for *Escherichia coli,* also known as colibacillus), has been the object of much of the seminal research leading to modern molecular biology.

Atmospheric oxygen was a major contribution of life to Earth

A second major event that took place during the "microbial era" is the appearance of molecular oxygen in the Earth's atmosphere. Oxygen, the life-giving gas par excellence, was absent in the early atmosphere, as attested by multiple evidence in ancient rocks. LUCA and its descendants for more than one billion years were all "anaerobic," which means that they lived without air, as many microorganisms still do today, for example, those that accomplish some of the fermentations on which we rely for the manufacturing of alcoholic beverages and cheeses. The early organisms, never having been exposed to oxygen, may even have been strict anaerobes, which are killed by this gas, as is the case for the bacillus of gangrene, which infects poorly aerated wounds and is readily killed by the simple device of incising the wounds and exposing them to air.

What introduced oxygen into the Earth's atmosphere was life itself, through the appearance of a special kind of photosynthetic bacteria called cyanobacteria (from the Greek *kyanos,* blue). These organisms use sunlight energy to split water (H_2O) into hydrogen (H_2) and oxygen (O_2). The hydrogen is used to convert CO_2 into sugar, from which, in turn, all the other organic cell constituents are formed, whereas oxygen is released in gaseous form.

According to the geochemical evidence, the atmospheric oxygen level started rising about 2.4 billion years ago, up to, first, a value of 1 percent of the terrestrial atmosphere, reached some 2.2 billion years ago. It stayed at that level until a second upward move lifted it to its present value of 21 percent of the atmosphere, about 1.6 billion years later.

The appearance of oxygen had a profound influence on the anaerobic forms of life that had been exclusively present until that time. Many disappeared, victims of oxygen toxicity, like the gangrene bacillus exposed to air. This extinction is sometimes referred to as the "oxygen holocaust," a misleading term that suggests a sudden catastrophe that took place almost overnight. In fact, the process was exceedingly slow, with the oxygen content of the atmosphere rising by much less than one-ten-thousandth of a percent per millennium. There was plenty of time for life to adapt to the changing conditions.

This adaptation probably took place first by the acquisition of enzymes capable of disabling oxygen. Such enzymes are present in all aero-tolerant organisms today. Eventually, some organisms not only protected themselves against oxygen but also acquired ways to take advantage of it. They developed chemical mechanisms whereby the energy released by the reaction of foodstuffs with oxygen could serve to support various kinds of biological work, much like fuel combustion supports motor-car engines and other heat-powered machines.

The biological machineries, however, are "cold" engines; they convert oxidation energy into work without the mediation of heat. Most of life today is powered entirely (animals, fungi, many bacteria) or partly (plants and other photosynthetic organisms in the dark) by such engines, to the point that oxygen, from being a deathly menace, has become an indispensable condition of survival for much of life on Earth. The most sophisticated biological "combustion engines" are found in certain bacteria and in mitochondria, about which more soon.

The birth of eukaryotic cells inaugurated a new living world

The third major biological event that happened in those early days was the epoch-making birth of a new type of cells of much larger size and more complex structural and functional organization than their bacterial predecessors, conspicuously including a central nucleus containing the genome. Called "eukaryotic" (which is Greek for "having a good nucleus"), as opposed to the prokaryotic bacteria, these cells gave rise to a wide variety of unicellular organisms, known as protists, and also to all multicellular organisms, including plants, fungi, animals, and humans. So the prokaryote-eukaryote transition represents a watershed in the history of life on Earth, a key event on the way to our own appearance. Without it, our planet would still harbor only bacteria.

Endosymbiosis was a key phenomenon in the development of eukaryotes

Eukaryotic cells are so different from prokaryotic cells that one finds it difficult to imagine how one type could ever have arisen from the other. Yet, this is undoubtedly what happened, given the many indisputable signs of kinship between the two. Although the problem is far from solved, a number of telling clues are already available. Particularly revealing is the astonishing fact, now solidly established, that two key, granule-shaped organelles (small organs) of eukaryotic cells were once free-living bacteria that, at some time in the distant past, were taken up by other cells within which they underwent a progressive process of enslavement, turning into "endosymbionts," (literally meaning "living together inside") and, eventually, evolving into fully integrated organelles.

First to be adopted in this way were the mitochondria, which are the main sites of oxidative energy production, the central "power plants," in the vast majority of eukaryotic cells. These organelles are derived from bacterial ancestors that must have ranked among the most efficient prokaryotic oxygen utilizers at the time they were adopted and have left similarly endowed, present-day descendants showing many signs of kinship with mitochondria.

The second eukaryotic organelles of established endosymbiont origin are the chloroplasts, which harbor the light-utilizing systems of all photosynthetic eukaryotic cells, to wit, all unicellular algae and green plants. The bacterial ancestors of these organelles have been identified as belonging to the group of cyanobacteria, encountered above as the "inventors" of oxygen-generating photosynthesis. These ancestral organisms were first adopted by cells that already possessed mitochondria, which are thus present in all photosynthetic eukaryotes (except when lost in the course of evolution).

For endosymbiosis to take place, there must first have existed cells with a size and functional properties that allowed them to harbor the bacteria that gave rise to the organelles. This question has been a fertile ground for all kinds of hypotheses, one more ingenious than the other. For my part, I stick to the simplest possibility, directly inspired by what we know and using a common cellular function, called "phagocytosis," whereby, for example, white blood cells capture infectious bacteria that invade an organism. We need merely to suppose that a "primitive phagocyte" possessing this property already existed at the time we are talking about and that, exceptionally, the bacterial ancestors of the endosymbionts captured by this organism were not killed and destroyed, as happens in white blood cells, but survived to become the endosymbionts. Such a phenomenon would hardly be surprising, as several present-day instances of it are known. According to the hypothesis I propose, this phenomenon would have happened at least twice, first to the ancestors of mitochondria and then, again, to the cyanobacteria that evolved inside the host cell to become the chloroplasts.

According to this scenario, formation of the "primitive phagocyte" from a prokaryotic ancestor appears as a crucial step in the development of eukaryotic cells. A detailed discussion of the manner in which this key transition could have occurred would take us too far. Let me simply emphasize the important role that may have been played by the passage from extracellular to intracellular digestion. All living beings that feed on nutrients provided by other living beings must start by digesting their foodstuffs, that is, cutting the big molecules of which these are made into small molecules that can be assimilated. This is what happens in our stomach and intestines. For single cells, this function is carried out in two different ways, depending on whether they are prokaryotic (bacteria) or eukaryotic. The former universally digest their foodstuffs with the help of enzymes that they discharge into their immediate surroundings, a process that requires prokaryotes to reside within their food source, like worms inside an apple or a piece of cheese. Eukaryotic cells, on the other hand, almost all feed by phagocytosis and digest their food within small intracellular pockets called "lysosomes"; they are thereby freed from the residential constraints to which bacteria are subjected. Thus, the development of the phagocytic mode of cellular feeding probably represents one of the key events in the birth of eukaryotic cells, the source of their emancipation and their ability to adopt endosymbionts.

Protists are the ultimate champions of unicellularity

We lack reliable fossil traces and thus do not know for sure when eukaryotic cells first arose. But we do know they have given rise to a multitude of unicellular organisms, or protists, which have gone on evolving up to the present day and exploiting the potentialities of unicellularity to their utmost, spreading into an extraordinary variety of organisms. These include the most elaborate and remarkable unicellular forms known, which have fascinated their observers by the multiplicity of their specializations, the elaborateness of their adaptations, and the beauty of their structures.

Multicellularity allowed division of labor

There is a limit, however, to what can be accomplished by a single cell, obliged to carry out all the functions needed for independent life. At some stage, the advantages of a "division of labor" must have favored the emergence of organisms genetically predisposed to form multicellular associations. Many mutually advantageous associations among members of the same or of different species no doubt formed, as they do today.

But true multicellular organisms were apparently late in appearing. Possibly accounting for this delay is the fact that true multicellular organisms are derived from a *single* egg cell, which gives rise to two or more distinct cell types by division and differentiation. Here is the key word: "differentiation." Starting with a single genome, different cells are generated by a process dependent on certain genes being expressed and others silenced, in a manner different for each cell type. Mechanisms for turning genes on and off are already present in the simplest of prokaryotes. But it probably took special circumstances to convert such primitive mechanisms into a developmental pattern. There will be more on this subject in chapter 6.

According to presently available evidence, multicellular forms of life appeared only about one billion years ago. Plants came first, soon followed by the fungi, or molds. Animals arose much later, about six hundred million years ago—that is, at the time when the atmospheric oxygen level went through its second rise, from 1 percent to 21 percent of the atmosphere. This is probably more than a coincidence, considering the absolute dependence of animals on oxygen. The three lines evolved in parallel, following comparable courses within the constraints imposed by their respective modes of life.

One common trend was a progressive rise in complexity, a quality that, to avoid the accusation of subjectivity and personal value judgment made by some philosophers, can be defined objectively by the number of different cell types of which organisms are made. This number increased from an original two to several tens in plants and fungi, and up to some 220 in animals. This rise in cellular diversity went together with increasingly elaborate arrangements of tissues and organs. Particularly intricate body plans were achieved in the animal line, with, among others, the appearance of neurons and their association into increasingly complex polyneuronal systems, of which the human brain is the most highly developed extant form.

Born in water, plants were the first multicellular organisms to invade land

A second feature common to the evolution of the three lines is that they all started in water and eventually invaded land, thanks to a variety of adaptations. Plants led the way, as they had to, since only they could do without other living organisms, being capable of constructing all their substance from water, carbon dioxide, and a few minerals, using light as energy source. The other two lines, being dependent, directly or indirectly, on the plants for food, could only invade land that had already been colonized by plants. I leave out here prokaryotes that could have served to feed very primitive forms of life.

Inaugurated in water by simple seaweeds, plants started to move out of their birthplace by way of coastal varieties periodically exposed to dryness at low tide and thus likely to benefit from traits favoring survival under dry conditions. These acquired attributes included rootlets capable of drawing water and minerals from the soil and coverings that both protected the plants against desiccation and allowed them to draw carbon dioxide from the surrounding air. Thus were born primitive mosses, the first multicellular organisms to invade land.

The mosses further evolved into the first vascularized plants, fitted with roots and leaves linked by a double set of conduits. One set of conduits, leading upward, served to bring to the leaves the water and mineral nutrients taken up from the soil by the roots. In the leaves, these nutrients were then combined with atmospheric carbon dioxide into various organic compounds with the help of sunlight energy. The other set of conduits served to convey the products of these syntheses from the leaves to the roots and other nonphotosynthetic parts, to be used for metabolism and growth. This basic design has been preserved in the entire further evolution of plants, leading, largely by way of improvements in reproductive strategies (see chapter 5), first to organisms represented today by ferns, then to organisms related to conifers, and, finally, to flowering plants, which make up much of the plant world today. An important development in this history was the "invention" of lignin, the hard substance of wood to which trees owe their remarkable strength.

The plants were soon followed on land by the fungi (mushrooms and molds), which, though being both dependent on other living organisms for their food supply and unable to move and hunt for food, have acquired the means to survive by utilizing whatever organic support, whether living or dead, they can stick to, deriving nutrients from it with the help of powerful digestive enzymes that they secrete in contact with their support.

The evolution of animals developed around the alimentary function

The story of animals is more complicated. Being obliged, like the fungi, to obtain their food from other living organisms, animals developed, like these organisms, around the indispensable functions of feeding and digestion, but in a different way. Their first ancestors, born in water, initially arose by exploiting the primeval phagocytic mechanism of feeding common to all protists. From first serving to support individual cells, as in sponges, this mechanism became communal in the digestive pouches of polyps and jellyfish, using enzymes secreted by the cells surrounding the pouch. Conversion of the pouch with a single opening—serving both for the entry of food and for the exit of waste—into a one-way canal, fitted with a mouth at one end and an anus at the other, completed the basic design of the animal alimentary tract, which has been maintained in all the forms that followed.

All other animal functions developed around this central alimentary core, in relation with the presence of cells that were increasingly distant from the digestive tract, while remaining dependent on it for their feeding. Thus were born circulation and, with it, respiration and excretion. Circulation served for bringing to the cells the foodstuffs and oxygen they needed and for clearing them of waste products. Respiration acted as a means, by way of gills and other organs, to capture oxygen and introduce it into the circulation for delivery to all cells. The function of excretion was to discharge, by organs such as kidneys, cellular waste carried by the circulation.

Another characteristic animal acquisition was motility, which was ensured by a variety of mechanisms, mostly dependent on the operation of special organs, the muscles. Organisms were thereby provided with all sorts of ways to seek food, find mates, join in groups, escape or fight predators, and so on. With motility came the neurons and the beginnings of a nervous system, serving first to adapt motile responses to sensory influxes and developing further into increasingly complex regulatory networks, thanks to the ability of neurons to establish connections (synapses) with each other. Chemical transmitters evolved as a means to use these connections to transmit signals from neuron to neuron, and these transmitters eventually developed into hormonal systems. Finally, all kinds of specializations were built around the all-important function of reproduction (see chapter 5).

Marine invertebrates inaugurated animal life

These events gave rise first to the rich world of marine invertebrates, which include the sponges and jellyfish already mentioned, corals, sea anemones, different kinds of worms, mollusks—characterized by a great variety of solid outer shells—arthropods, such as lobsters, crabs, and other crustaceans—distinguished by an articulated outer skeleton made of a very tough substance called chitin—and, characterized by a peculiar fivefold symmetry, echinoderms, of which starfish and sea urchins are the best-known representatives.

Body segmentation opened the way to vertebrates

A key event that occurred at some early stage of this development was the repeated duplication of a central set of genes (see chapter 6), which led to *segmentation,* the building of bodies made of a large number of similar units. Almost identical at first, as in the familiar earthworms, these units later evolved into a wide variety, illustrated, for example, by the antennae, claws, and other appendages of crustaceans. Eventually, the units produced the characteristic segments of vertebrates, starting with primitive fish, which further evolved into more advanced fish and, from these, into all the forms that followed.

Several distinct animal lineages moved from water to land

Adaptation of animals to living on land involved several key acquisitions: a skin capable of protecting against desiccation, a mechanism for deriving oxygen from air instead of from water, and a motor system allowing movement on land. Ability to reproduce on land, as we shall see in chapter 5, was another essential requisite. Remarkably, several distinct such adaptations developed at different stages of animal evolution. For example, marine worms turned into nematodes and, in a later, segmented line, into earthworms; aquatic mollusks evolved into snails; and arthropods gave rise to the vast group of insects and arachnids (spiders, scorpions, and the like). As to vertebrates, their transition from water to land probably took place in shallow tropical lakes that periodically evaporated during the dry season and regained water during the rainy season. Some fish, known as lungfish, of which species still exist today, became able to survive on land thanks to a dryness-resistant skin, rudimentary lungs derived from the swim bladder, and modified fins converted into primitive limbs. Thus arose, some 400 million years ago, the first amphibians, represented today by animals such as frogs, salamanders, and toads, which still depend on water for their early development. Then, about 350 million years ago, some amphibians evolved into the first vertebrates fully adapted to live and reproduce on land, the reptiles, made famous by the giant dinosaurs, which fill museums with their spectacular remains and have inspired innumerable works of fiction.

Dinosaurs gave rise to birds and mammals

Further vertebrate evolution took place on land. Some dinosaurs acquired feathers, perhaps serving initially as a protection against a cold climate, and eventually turning into primitive wings that allowed the animals to glide and, later, to fly. First revealed by archaeopteryx, the fossil of a feathered, presumably flying dinosaur, discovered in 1864 in a Bavarian schist quarry, this story has since received confirmation from a number of fossils found in China. Its outcome is the appearance of birds, about 150 million years ago.

Other dinosaurs became covered with hair and acquired milk-secreting glands on their chest, allowing females to feed their young. This acquisition led, some 225 million years ago, to the first mammals. These creatures remained small, enjoying a relatively modest existence in the shadow of the monstrous dinosaurs, until some 65 million years ago, when a planetary catastrophe, probably initiated by the fall of a large meteorite on the Yucatán Peninsula in Mexico, precipitated the massive extinction of dinosaurs and many other animal and plant species. Subsequent to this cataclysm, mammals underwent an extraordinary development and came to occupy all environments, even returning to the sea in some cases, as happened to the ancestors of seals and whales. Mammals gave rise, some 70 million years ago, to the primate group, out of which a line detached, some 6–7 million years ago, that was to lead to the human species.

Viewing this grand history (fig. 3.1), or rather its present outcome, through the eyes of the prophets who wrote the Bible or of medieval scholars, who didn't even know about microbes, one can readily understand how this whole pageantry was viewed as given once and for all, brought into being by a Creator for the sole benefit of humankind. Even the eighteenth-century Swedish naturalist Carl von Linné (1707–1778), who did know about microbes and who spent his entire career observing and describing living organisms, patiently classifying them into species, genera, families, orders, classes, phyla, and kingdoms, failed to see that the kinships he was recognizing rested, like those of human families, on a vast genealogical tree springing from a single root. Linné remained all his life an unconditional defender of "fixism" and adhered staunchly to

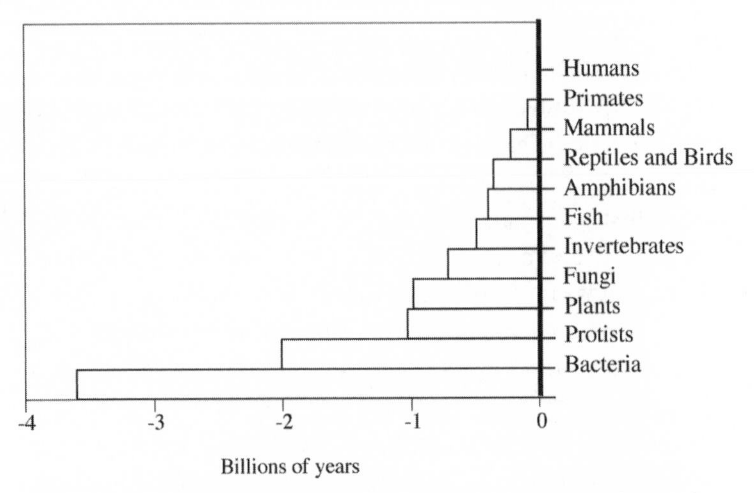

Fig. 3.1. *The main steps in the history of life, in particular of animals.* Note that life remained exclusively unicellular during 2.5 billion years. The first animals appeared 600 million years ago, after life had already accomplished five-sixths of its history. The human species dates back a mere 200,000 years, the equivalent of the last half-hour if life had started one year earlier (and animals two months earlier).

the biblical story. Even his later French successor Georges Cuvier (1769–1832), the founder of comparative anatomy and paleontology, adamantly refused to accept the transformist hypothesis proposed by his rival Lamarck, even though he was hardly influenced by biblical creationism. We don't have their excuses today. Evolution, as we have seen, no longer calls for demonstration.

II
The Mechanisms of Life

Introduction

In the first part of this book, I sketched a descriptive picture of the main steps of evolution. But describing is not enough for understanding. One must explain. All historians know this. That is what I try to do in this second part. It starts with three chapters devoted to three fundamental biological mechanisms that need to be known by anybody wishing to understand life and its history: metabolism, the entire set of chemical reactions that underpin the functioning of living beings since their first appearance; reproduction, which has served as a link between generations all along evolution and ensures hereditary transmission; and, finally, development, which covers the processes whereby, in multicellular beings a fertilized egg gives rise to an organism.

Natural selection, the topic of chapter 7, represents the central theme of the book, the conducting thread between past and future that leads to the warning at the end of this book, the beacon that illuminates the entire history of life, up to its most recent steps and, even, its future prospects.

There will be a brief mention, in the last chapter of this part, of some of the other evolutionary mechanisms that have been proposed, including "intelligent design," which is not properly speaking a scientific mechanism, but warrants attention because of the media upheaval it generates.

4
Metabolism

The *New International Webster's Comprehensive Dictionary* defines "metabolism" as "the aggregate of all physical and chemical processes constantly taking place in living organisms." The key words are "physical" and, especially, "chemical." There is no escape. If one wishes to understand life, one has to go through some chemistry. In a book like this, we can't examine all the details that fill biochemistry textbooks with formulas of daunting complexity. Fortunately, it is possible to give an idea of how metabolism works without calling on a single formula. This is what I try to do.

Living cells are chemical factories

Have you ever visited a chemical factory? If you've seen one, you've seen them all, for all are constructed on the same model: a collection of closed vats linked by pipes. Each vat is the site, under specified conditions of temperature, acidity, and so on, and with the eventual addition of a catalyst to facilitate the reaction, of a given step in the specified process. The pipes feed reactants into the vat and allow exit of the products. Raw materials are introduced into the system. They circulate from vat to vat, while undergoing progressive transformations, finally to exit as finished products. The pathways thus followed vary with the nature of what is manufactured. They can be more or less complicated but rarely comprise more than a few tens of steps.

Living chemical factories follow the same model, except that they carry out a large number of different production programs simultaneously, that they include many more steps, and that, aside from a few compartments, such as mitochondria, that house a large number of reactions, there are no vats and no pipes, or their equivalent. It all takes place in a single phase, or *metabolic pool,* containing all the participating substances. This is possible because these substances may rub each other without in the least interacting. The circulation of matter through the system is entirely ensured by the catalysts of the reactions, most often *enzymes* of protein nature.

Enzymes display on their surface binding sites that specifically fish out of the metabolic pool the substances, or substrates, that participate in the reaction catalyzed by the enzyme. The substances thus caught find themselves within the field of action of another site, called catalytic site, that ensures their transformation. Once the reaction is terminated, its products fall off the enzyme and join the metabolic pool.

Foodstuffs brought into the cells from the outside circulate from enzyme to enzyme in such systems, progressively transforming into the final products. These include: cellular constituents, made to replace damaged molecules and to satisfy the needs of growth, reserve substances that are held in storage by the cells, and, to be discharged outside, eventual secretory products and waste substances. The pathways followed by this chemical circulation are called metabolic pathways.

Living cells extract the energy they need from their surroundings

Cellular factories, like chemical factories, require energy to support their activities. Many different mechanisms have evolved to generate this energy in relation with locally available sources. There are *heterotrophic* organisms and *autotrophic* organisms. The former derive their energy from the degradation, with oxygen (aerobic) or without it (anaerobic), of organic foodstuffs provided by other (*heteros,* in Greek) living beings. They use the same foodstuffs as building blocks for their biosyntheses. Such is the case for all animals, including humans, and for fungi and many microbes.

Autotrophs are divided into *photosynthetic,* which derive energy from sunlight, and *chemosynthetic,* which exploit mineral chemical reactions. Green plants and algae belong to the first group. Methanogens (chapter 3) are a particularly simple example of the latter. Autotrophs differ from heterotrophs by their ability to do without any foodstuff of living origin for their biosyntheses. Hence their name, which underlines the fact that they are self-sufficient (*autos* means self in Greek); they have no need for any other living organism. They use water, carbon dioxide, sometimes atmospheric nitrogen, and a few mineral elements that they extract from the soil. Their foodstuffs, when needed, are the mineral fertilizers, with, among others, nitrates as source of nitrogen, that gardeners or farmers provide when the soil is too poor.

Remarkably, this extraordinary diversity of mechanisms clusters around a bioenergetic core common to all living organisms and centered on a key compound designated by the acronym ATP (for adenosine triphosphate). This substance also serves, with the help of appropriate transformers, as source of energy for the other forms of work—motor, electric, osmotic, informatic, and so on—carried out by living beings.

ATP is the *universal energy mediator.* It is sometimes replaced in that capacity by closely related chemical substances known as GTP (guanosine triphosphate), CTP (cytidine triphosphate), and UTP (uridine triphosphate). The four "NTPs" (nucleoside triphosphates) are also the basic precursors in the synthesis of ATP. One recognizes the four canonic bases—A, G, C, and U—already mentioned in the first chapter. This fact creates a bridge, of impressive significance for the origin of life, between energy and information.

Thousands of specific catalysts are involved in metabolic reactions

Metabolic pathways are delineated by the agents that catalyze them. These comprise mainly the protein enzymes, already mentioned. A few natural catalysts are of RNA nature and are called *ribozymes.* To this catalytic armamentarium must be added a number of small molecules, called *coenzymes,* that, as their name indicates, play an essential auxiliary role in many enzymatic reactions. Coenzymes often contain a vitamin as active constituent. Several of them include in their structure a derivative of one of the NTPs mentioned above.

In turn, enzymes and ribozymes are synthesized, together with other cellular proteins and RNAs, according to blueprints stored in DNA molecules, subject themselves to replication, all of it being catalyzed, like everything that goes on in cells, by specific enzymes and ribozymes.

Metabolic pathways form networks of enormous complexity

Most of the substances that participate in metabolism are involved in a dual capacity, as products of one or more reactions and as substrates (reactants) of one or more others. Those substances, called *metabolic intermediates,* or *intermediary metabolites,* link together the reactions concerned. As an example, imagine two reactions: one whereby substance A is converted into substance B, and another that converts B to C. Those two reactions are linked together by the intermediate B, product of the first reaction and substrate of the second: A→B→C. This is the start of a linear pathway that could be prolonged by reactions in which C leads to D, D to E, and so on. Things can be more complicated.

Thus, if a second reaction starting from B exists, leading to C′, B becomes the origin of a bifurcation of which one branch leads to C and the other to C′. Things can be even more complicated, with, for example, substances other than A converging on B, or with reactions involving two different substrates issued from two different pathways, as is the case for most metabolic reactions, or again with intermediates participating in more than two reactions, and so forth. Such assemblages can lead to a vast, multidimensional network made of linear pathways, bifurcations, crossroads, stars, roundabouts, cycles, and even more complex configurations.

Cell metabolism constitutes such a network. Represented by the metabolic map, it is a single network in which everything holds together, with a few rare entrances for outside substances feeding into the network, and a number of exits leading newly synthesized cell constituents to their loca-

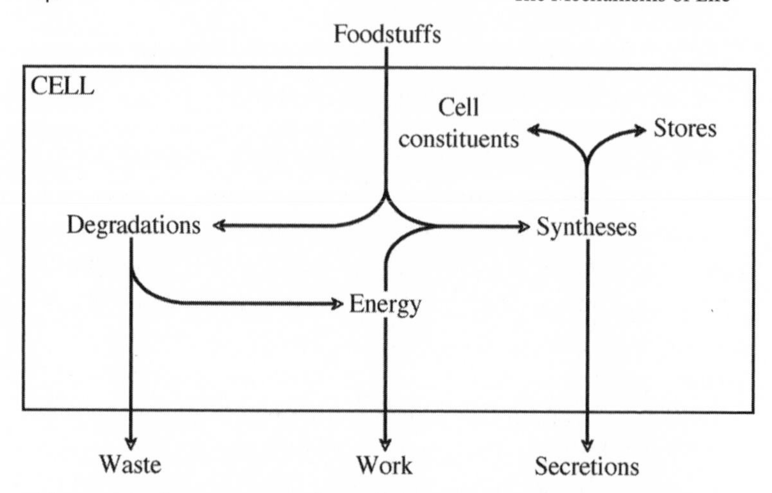

Fig. 4.1. *A schematic view of metabolism.* Foodstuffs provided from outside enter the metabolic network, where they are either degraded with the production of energy or used for syntheses. The energy freed by the degradations serves to sustain the syntheses, as well as the other forms of work carried out by the cell. Synthetic processes serve to form new cell constituents to satisfy the needs of growth and repair, reserve substances, which are stored, and secretion products, which are discharged outside, together with metabolic waste products.

tions in the cells, reserve substances toward their deposit sites, and waste products and secretory materials to the outside (fig. 4.1). Think of the road map of a country, with its limited entry and exit points at the borders. The complexity of the metabolic network, however, exceeds by far that of our densest roadway networks. Some of its crossroads, such as those occupied by coenzymes that participate in up to several tens of reactions, may form the starting and endpoints of as many distinct roads. L'Étoile in Paris, Piccadilly Circus in London, or Times Square in New York pale by comparison.

It is a dynamic, perpetually changing network, in which

the circulation of matter is subject to an equally complex set of automatic regulations that constantly adapt to each other the velocities of the reactions involved. In it, certain spots are the targets of outside influences that allow the network to adjust to changes in the milieu or to respond to hormonal or nervous messages. Many poisons and drugs act by intervening at one or the other site of the metabolic network, by inhibiting a reaction, for example, causing jams that may paralyze the entire network.

Metabolic networks vary with cell types, which owe their particular properties to those networks. But they include certain central pathways that are common to a vast majority of cells, such as those known by experts under the names of glycolytic chain or of Krebs cycle, or the systems of protein synthesis. These pathways probably go back to the beginnings of life.

We are what our catalysts are

This statement summarizes the fact that all we accomplish depends on chemical reactions and that these depend entirely on the enzymes, ribozymes, and coenzymes we possess. There are innumerable proofs of this.

Take a substance such as vitamin PP (pellagra preventing), or nicotinamide. It is a very simple, small molecule, which must have appeared very early in the history of life, as it is a key constituent of two central coenzymes, called NAD and NADP by biochemists, present in the near-totality of living beings, where they play an essential role in many metabolic reactions. In the course of evolution, one of our distant ancestors lost an enzyme required for the synthesis of this substance, bequeathing this defect to us. This is the reason why we must find nicotinamide in our food. Such is the case for all vitamins.

Otherwise, they wouldn't be vitamins. For some, the loss of a critical enzyme is relatively recent. Thus, primates (to which we belong) and guinea-pigs are the only animals subject to scurvy, the illness caused by a deficiency in vitamin C. All other animals manufacture their own vitamin C. It is told that long-distance voyagers suffering from scurvy because of lack of fresh food recovered from the illness when they were reduced to eating the ship's rats.

The loss of an important enzyme may even be more recent. Many congenital diseases are due to the genetic deficiency of an enzyme and were called "inborn errors of metabolism" by Sir Archibald Garrod, the English pediatrician who first discovered such a condition. Those diseases most often affect a small number of families. Therefore, the mutations responsible for them must have happened in almost contemporary individuals. Tay-Sachs disease, for example, which is restricted to certain Jewish populations, is due to a mutation that has been traced back to the Middle Ages in a small central-European village.

Before losing enzymes, living beings must first have acquired them. Life obviously did not arise with a full set of thousands of specific biocatalysts. This would have required instantaneous creation.

The history of metabolism goes back to the earliest days of life

The history of metabolism coincides with that of the biocatalysts and can only have been progressive. Sticking to protein enzymes, which catalyze the vast majority of metabolic reactions, molecular studies have allowed two important evolutionary mechanisms to be recognized.

First there is gene *duplication,* with both copies being retained in the genome. This phenomenon allows evolution to "tinker"—the imaginative wording coined by the French biologist François Jacob—with one copy of the gene, eventually making something new out of it, while conserving the other copy unchanged to keep its function going. Numerous examples of this fundamental mechanism are known.

Another important mechanism is *modular combination.* It has been found that many proteins—and, therefore, the corresponding genes—are composed of a number of distinct blocks endowed with special functions, such as a given catalytic activity or the ability to bind a certain substance, a coenzyme, for example. The same blocks are found, often diversely associated, in different protein molecules, suggesting that these are the products of a combinatorial game involving a limited number of modules.

The existence of those modules throws an interesting light on the beginnings of proteins, by suggesting that these substances started in the form of very short chains. This suggestion agrees with theoretical studies that lead, by a totally independent argument, to the conclusion that the first genes must have been very short. It is conceivable that the first enzymes were present among the translation products of those genes, displaying catalytic functions that were no doubt rudimentary, but sufficient to play a role in nascent metabolism. Time, mutations, duplications, tinkering, modular combinations, and natural selection (see chapter 7) have done the rest, finally creating the network of mind-boggling complexity, consisting of thousands of highly sophisticated catalysts, that underlies today's metabolism.

But this is not all. Before proteins, as we saw in chapter 2, there were RNAs, which are credited with the "invention" of proteins. Hence the hypothesis, proposed more than twenty years ago, that the stage that preceded metabolism catalyzed by protein enzymes was activated by ribozymes. This notion has been enormously successful, under the appellation of "RNA world," a hypothetical stage in the origin of life, in which RNA molecules are taken to have played the role of catalysts of the first metabolic reactions and, at the same time, also the role of replicable repository of genetic information (RNA, as we have seen, also preceded DNA).

Without entering into the enormous array of discussions and experiments engendered by this model, I simply wish to underline that it fails to explain certain fundamental questions, including, most important, the origin of RNA itself, the Holy Grail of research on the origin of life (chapter 2). It is my opinion, in agreement with that of a number of investigators, but against that of the most enthusiastic defenders of a "pure" RNA world, that a complex chemical infrastructure must have been required to inaugurate this stage in the development of life and to sustain it during all the time it took RNA molecules to generate the first proteins able to assist them by their catalytic activities. This "protometabolism," as I call it, could already have included certain key reactions of present-day metabolism, issued from protometabolism in congruent fashion. This view is not shared by a number of experts, who believe that prebiotic chemistry was very different from biochemistry.

As to the indispensable catalysts required by protometabolism, suggestions are that they could have been minerals, such as clays; or organic compounds such as peptides, substances similar to proteins, but that could readily have formed under prebiotic conditions; or, again, self-supporting autocatalytic circuits, that is, chemical circuits that generate their own catalysts. For my part, I lean in favor of peptides, to which I add analogous substances, under the common appellation of "multimers." Some key coenzymes could already have participated as well in this protometabolism.

The stage at which ATP and its homologues, GTP, CTP, and UTP, first arose raises an intriguing problem. We have seen that these compounds play a dual role of key importance in present-day life: in energy transfer and, as precursors of RNA, in information transfer. The question is: Which of these two functions did the compounds serve first, and how? An implicit feature of the RNA world hypothesis is that information came first (with RNA). My own view is that the building blocks of RNA, that is, the NTPs, must have come first, perhaps inaugurated by ATP, thus accounting for the central role of this compound in metabolism throughout the living world. Supporting this contention is the fact, admittedly only negative, that no pathway to RNA circumventing the NTPs has been discovered or, even, imagined so far. This key question has not yet been subjected to an experimental test that could either confirm or disprove one or the other hypothesis. Time will tell.

5
Reproduction

Reproduction is a fundamental property of life, the driving force of life's continuity, generation after generation, from the time of its first appearance up to present-day living beings.

Reproduction started with molecular replication

A fundamental property of molecular replication is that it does not rely on direct copying, as in an office copier, but on *complementarity*, as in photography, with a negative serving to assemble a positive, and vice versa. This key mechanism, which was most likely inaugurated by RNA in the origin of life, was discovered first for RNA's better known sister molecule, DNA, by the American James D. Watson and an Englishman, the late Francis Crick, who, in 1953 published their historic paper describing the double-helical structure of the DNA molecule, perhaps the greatest discovery ever made in the life sciences.

The novel notion in this historic proposal was not so

much the helical shape of the DNA molecule, which is imposed by the angles of the chemical bonds and had been suspected before, but rather the molecule's *double* character and, especially, the physical basis of this duality. DNA consists of two strands, twisted together like the threads of a string or, as a more appropriate image, the two sides of a spiral staircase (fig. 5.1). Each of these strands consists of a continuous thread, which is the same for all DNA molecules and is made of alternating molecules of phosphate and of the sugar deoxyribose (hence the name of *deoxy*ribonucleic acid, or *D*NA). To this thread are attached, like flaglets to a string, small, nitrogenous molecules, called bases, belonging to four different kinds, which we shall represent simply by their initials: A, G, C, and T. As already seen in the first chapter, the sequence of bases specifies the information content of the molecule, a veritable molecular "word" written with an "alphabet" of four "letters."

In the DNA double helix, the two strands are complementary according to a very simple rule, called *base pairing:* A in one strand always faces a T in the other strand, and G in one strand always faces a C in the other strand. This means that, with the sequence on one strand known, the other can be inferred. As a simple example, sequence GCCTAT on one strand automatically requires CGGATA on the other. This fundamental property is ensured by the molecular structures of the complementary bases, which have the shape of small flat pieces that fit into each other like two pieces of a puzzle and "stick" together by weak chemical bonds called hydrogen bridges. In the double helix, the base pairs thus formed follow each other like the steps of a spiral staircase, with the threads to which they are attached making the outer frame of the staircase (see fig. 5.1). Structures formed in this way may contain thousands, if not millions, of base pairs and reach lengths of

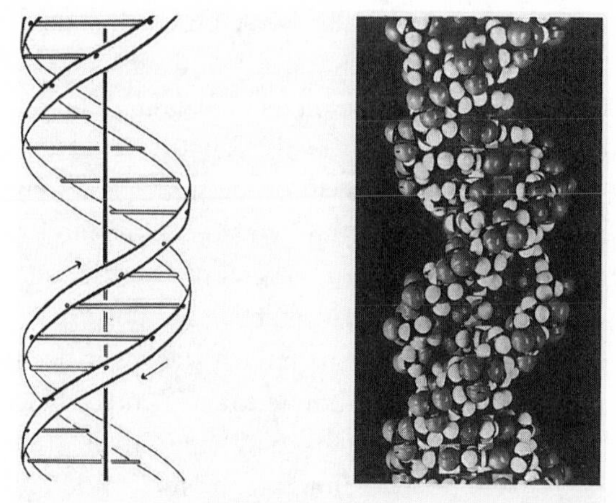

Fig. 5.1. *The double helix.* Left, schematic model, as it appeared in James D. Watson and Francis Crick's original 1953 paper. The spiral-staircase structure is clearly visible. Right, space-filling molecular model constructed by Maurice Wilkins, who shared the 1962 Nobel Prize in medicine with Watson and Crick. Left, Reprinted by permission from Macmillan Publishers, Ltd. Right, Courtesy of Maurice Wilkins.

an inch or more. They make up the genome, which consists of thousands of units called genes, each of which contains at least several hundred base pairs.

Base pairing does not only account for DNA's duplex structure, but also for its replication. The chemical system that assembles a new DNA chain using an old one as template automatically inserts T in front of A in the template, and vice versa, G and C likewise calling for each other (fig. 5.2).

A very similar base-pairing code rules the replication of RNA, which closely resembles DNA, with the only chemical differences being that deoxyribose is replaced in the common thread by ribose (hence the name of *ribo*nucleic acid, or *R*NA) and the base T is replaced by U, a close relative that, like it, pairs with A. In addition, in nature, RNA rarely exists in double-

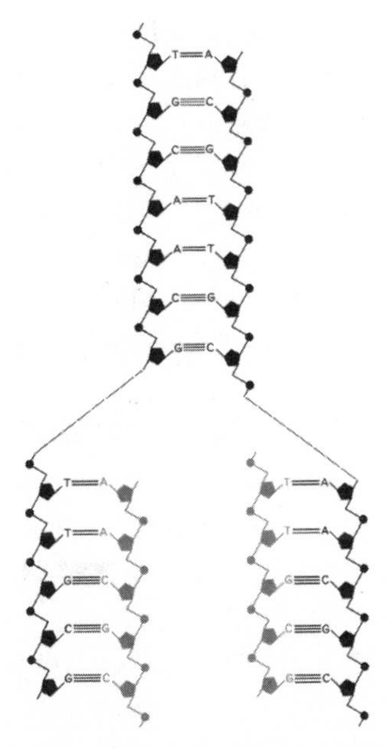

Fig. 5.2. *DNA replication.* The top half shows part of a DNA double helix, unwound so as to expose the joining of the two complementary strands by base pairing: A joins with T by means of two hydrogen bonds; G joins with C by means of three hydrogen bonds. The bottom half pictures synthesis, guided by the same base-pairing mechanism, of two new complementary strands on the separated strands of the original DNA. (Knowledgeable readers will note that this schematic diagram by the discoverer of DNA replication leaves out the fact, discovered after the picture was drawn, that the two strands are replicated in reverse directions. After A. Kornberg, in *Molecules to Living Cells: Readings from "Scientific American"* (New York: W. H. Freeman, 1980), 270–280.

helix form. It is usually made of a single thread, most often folded into a tangle of loops closed by short double-helical joints linking complementary stretches situated at distinct sites of the same thread.

It is very probable, as we have seen in chapter 2, that, in

the origin of life, RNA preceded DNA as replicable bearer of information and, therefore, that replication developed first for RNA. Today, this ancestral phenomenon takes place only in cells infected by certain viruses (the polio virus, for example) that possess an RNA genome. Everywhere else, replication concerns DNA. Historically, however, RNA replication was probably the first manifestation of base pairing, inaugurating what may well be the most fundamental process in the whole history of life on Earth.

Indeed, base pairing has turned out to be the dominant mechanism for information transfer throughout the living world, from the origin of life to the present day. It does not just rule DNA and RNA replication, but also the transcription of DNA into RNA and the opposed process of reverse transcription, the synthesis of DNA on an RNA template, which is carried out by certain viruses, for example the causal agent of AIDS. Base pairing also plays a fundamental role in the many interactions between RNA molecules that take place in the translation from RNA into proteins and in many other processes. It is the key mechanism in the universal language of life.

With the appearance of cells, cell division was added to molecular replication in biological reproduction

Reproduction remained molecular until the appearance of the first cells. After that, DNA replication had to be followed by doubling of the cells that contained the DNA, so that each daughter cell would be left with one of the two DNA copies. This doubling first occurred by simple division and, later, in eukaryotic cells, by a much more complex process called mitosis. We won't go into the details of this process, except to note that it involves rodlike structures, called chromosomes, bearing the DNA molecules that make up the cell's genome. Each cell division is preceded by duplication of the chromosomes, itself intimately linked with replication of their DNA content.

Multicellular beings reproduce by way of single mother cells

Such mechanisms sufficed as long as organisms remained unicellular. Once the first multicellular organisms appeared, a new reproduction mechanism evolved. Barring some rare exceptions, such as the reproduction of certain plants by budding, all multicellular organisms originate from a single mother cell that, by division and differentiation, gives rise to all the cells of the organism, and is called *totipotential* for that reason. It might be assumed, a priori, that this mother cell would arise in a parental organism, either from a differentiated cell returning to the totipotential state by "dedifferentiation" (fig. 5.3), or, as proposed by the German biologist August Weismann (1834–1914), from a continuous line of totipotential cells dividing asymmetrically to give rise, on one hand, to a totipotential cell that perpetuates the line, called "germ plasm" by Weismann, and, on the other, to a cell committed toward the formation of differentiated cells and eventually leading to the new organism (fig. 5.4).

Such mechanisms are not involved in the reproduction of organisms, but they play a role in other phenomena of considerable interest. Thus, the Weismann hypothesis accounts for many cases of cell renewal. In the bone marrow, for example, the various blood cells arise from a continuous line of so-called *stem cells,* which divide asymmetrically to give one daughter cell destined to differentiate further into a red blood cell or one of the various types of white blood cells, while the other daughter cell remains a stem cell. Similar processes take

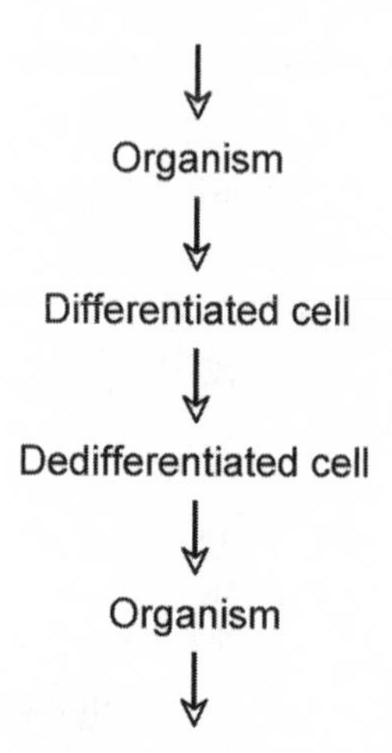

Fig. 5.3. *Hypothetical model of reproduction from a somatic differentiated cell that dedifferentiates into a totipotential cell leading to a new organism.* This phenomenon is not involved in the reproduction of organisms, but it is, to a certain extent, in cancerous transformation and, especially, in artificial cloning (see chapter 15).

place in most other organs, thereby replacing damaged cells. Even brain cells, which had long been seen as irreplaceable, can be generated by this mechanism. The possible therapeutic use of such "somatic" stem cells (from the Greek *soma,* body) for tissue repair has evoked enormous interest in recent years, especially because their use does not encounter the same ethical objections as does the use of embryonic cells, which is condemned by a number of religious groups because it involves the destruction of a potential human being.

As to dedifferentiation, it occurs, for example, in the conversion of normal cells into cancer cells, which are thereby almost returned to the status of rapidly dividing embryonic cells. Dedifferentiation has also become a subject of burning interest in relation with artificial cloning techniques. Recently, headlines were made by the announcement that certain dif-

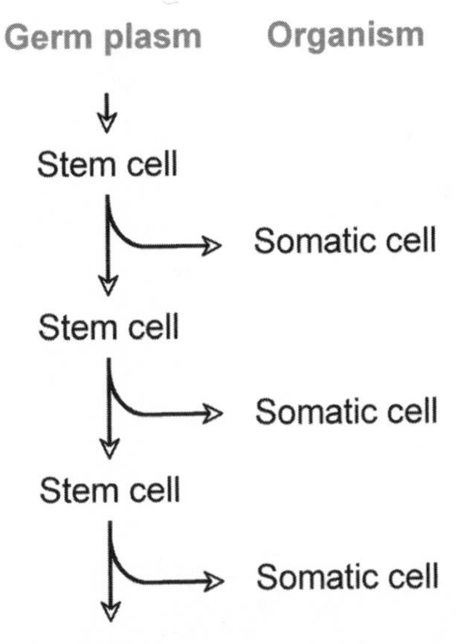

Fig. 5.4. *Weismann's theory.* August Weismann postulated a continuous germ line from which successive generations of organisms branch out laterally by asymmetric division. The model does not apply to the reproduction of organisms but accounts for the formation of somatic cells from pluripotential stem cells.

ferentiated cells can be induced by relatively simple means to return to stem cell status, another potential breakthrough in the production of stem cells for therapeutic purposes. We shall return to these important issues at the end of the book (see chapter 15).

The mother cell of multicellular beings arises from two parental cells by sexual reproduction

The mechanism almost universally used for reproduction by multicellular organisms involves, not one, but *two* cells. It is sexual reproduction (fig. 5.5). In this process, the mother cell from which a new organism is destined to arise is the product

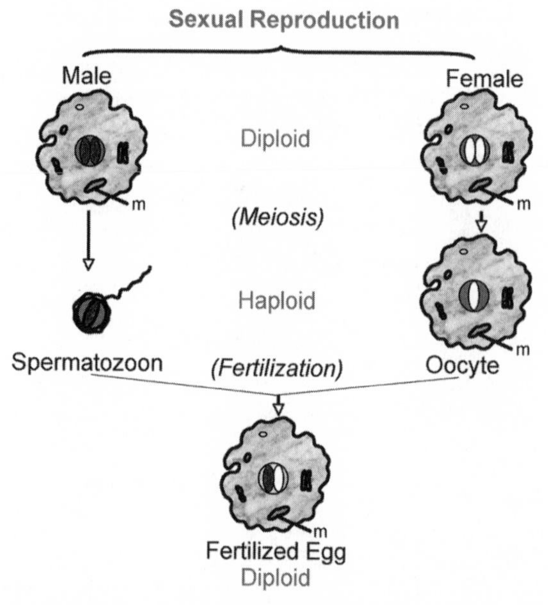

Fig. 5.5. *Sexual reproduction.* This diagram illustrates the maturation, by way of meiosis, of haploid male and female gametes from diploid mother cells, and the formation of a diploid fertilized egg by fertilization of the female oocyte by the male spermatozoon. Note that cytoplasmic organelles, including mitochondria, are eliminated in the course of sperm maturation, but are conserved in the course of oocyte maturation. This phenomenon is taken advantage of in the phylogenetic procedure based on the comparative sequencing of mitochondrial DNA (see mitochondrial Eve, chapter 9).

of the fusion of two distinct cells, most often with very different properties. In technical jargon, these cells are called "gametes" or "germ cells," their properties are distinguished by the terms "male" and "female," their fusion is known as "fertilization," and the product of this process is called a "fertilized egg cell."

Chromosome doubling caused by sexual reproduction is corrected by meiosis during gamete maturation

One wonders how sexual reproduction can ever have developed, as it implies a phenomenon that, according to every prediction, should have had a lethal effect, namely the multiplication of chromosomes, whose number doubles with every generation due to the fusion of two cells. This drawback was eluded by the development of a special kind of mitotic division, called *meiosis,* in which the double, or *diploid,* number of chromosomes inherited from the fertilized egg is reduced back to a single, or *haploid,* set in the course of germ-cell maturation. Thus, when two (haploid) gametes, male and female, join in fertilization, they generate a (diploid) fertilized egg, containing two sets of chromosomes. All the cells of the organism that arise through the development of the egg are likewise diploid, with the exception of the cells destined to become gametes. These cells undergo meiosis in the course of their maturation and become haploid, ready to repeat the cycle. This alternation between haploidy and diploidy is called *alternation of generations.*

Surprisingly, sexual reproduction, with its attendant passage through meiosis, occurs in the three multicellular lineages, plants, fungi, and animals. Development of such a complicated mechanism independently three times defies plausibility. One is thereby led to look for its origin in protists. Unicellular eukaryotes do indeed sometimes engage in this kind of reproductive fusion, especially under conditions of stress. Even prokaryotes occasionally practice what is known as conjugation, a process in the course of which two such cells exchange genetic material, thus creating new genetic combinations.

Sexual reproduction is the laboratory of evolution

Here probably lies the main advantage of cell fusion. It offers opportunities for testing new combinations of genes, which may be a vital asset when genetic innovation becomes a crucial condition of survival. This is all the more true because it is during meiosis that the process called *crossing-over,* or recombination, takes place. In this process, pieces are exchanged between chromosomes of the same pair, thereby creating unique combinations of genetic material that were not present before and offering evolution an almost infinite variety of genetic motifs to play with.

Sexual reproduction represents the veritable *laboratory of evolution.* Thanks to it, innumerable genetic variants have been continually subjected to screening by natural selection (see chapter 7). Genesis of this mechanism no doubt constitutes a key step in the development of multicellular organisms. The complexity of this step perhaps explains why multicellular life, as we know it, was so late in appearing.

Male and female gametes differ

A feature of sexual reproduction common to the vast majority of plants and animals is the participation of two distinct types of germ cells with very different properties and functional roles. The female germ cells, or *oocytes,* sometimes also called (unfertilized) egg cells, are large and immobile, fitted with a full complement of cytoplasmic structures and crammed with abundant nutrient reserves and other essential substances. The role of the female germ cell is to passively await fertilization and then provide all that will be needed to start development. In contrast, the male cells, called *spermatozoa,* or sperm cells, are small and motile, reduced to little more than a nucleus devoid of surrounding cytoplasm and propelled by an undulating tail, or flagellum. Their function is to seek a compatible egg cell and penetrate it, or, rather, insert their nucleus into it, which is all that is needed to convert the haploid egg cell into a diploid fertilized egg. A significant consequence of this mechanism is that the mitochondria of the fertilized egg are exclusively derived from the female germ cell (see fig. 5.5). This property is exploited in the analytical method used to trace descent by the female line (see chapter 9, mitochondrial Eve).

Relative to this difference in functions, male gametes are always produced in large numbers, and female gametes in very small numbers. This division of labor is energetically economical, as a female gamete is much costlier to make than a male gamete. This leaves to the sole male gametes the task, favored by their large number, of seeking a female gamete to fertilize. Many specializations of the corresponding organisms are related to the different functions of the gametes they produce (see below: *sexual dimorphism*).

Plant reproduction involves spores

Reproductive strategies have evolved very differently in plants and animals. In the latter, the haploid stage in the alternation of generations is invariably fleeting and transient, leading almost directly from meiosis to gametes through a short succession of cell divisions (maturation) that takes place in the sex glands of the male and female organisms, whereas the rest of the bodies of each sex consists entirely of diploid cells. In plants, the pathway from meiosis to gametes goes by way of an intermediate, haploid form, called a *spore*, which undergoes a variable degree of development, sometimes very complex, before giving rise to the gametes (fig. 5.6). This haploid stage may

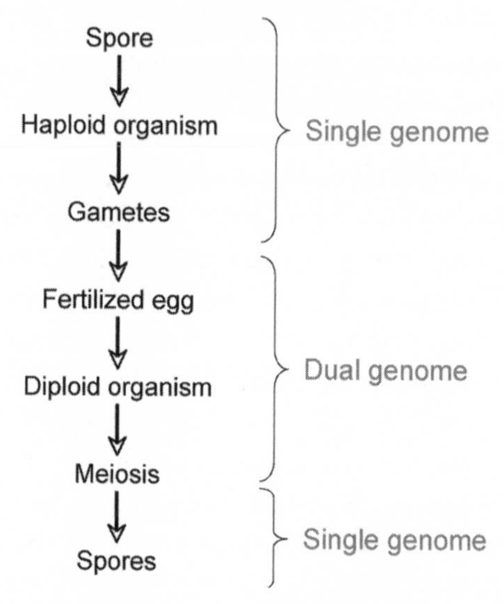

Fig. 5.6. *Alternation of generations in plants.* This diagram shows how plant life alternates between haploid and diploid forms, by way of spores, on one hand, and of sexual reproduction, on the other. The relative importance of the two forms varies according to the type of organism. In animals, the haploid phase is reduced to the maturation of the gametes following meiosis.

be up to dominant in certain cases, while the diplod stage plays only a brief, transient role.

The primitive seaweeds illustrate this situation in exemplary fashion. In some species, the haploid and diploid stages have the same importance and may even be almost identical in appearance. In others, either one or the other stage is dominant, with the other stage serving only a transitional role. Mosses, the first land-adapted plants, are largely haploid and rely on a brief diploid stage to move from one haploid generation to the other. In the more evolved ferns, the situation is reversed; the main form is diploid, and the haploid stage is only a short interlude occurring underground.

This difference between animals and plants is linked to the very different life-styles of the two types of organisms, which impose different strategies to allow the indispensable encounter between the gametes of the two sexes. Animals take advantage of their mobility to ensure this encounter. The immobile plants, on the other hand, rely mostly on external agents, such as wind or, alternatively, insects or other animals. Hence the need of a transportable form of gametes. Such is the role of spores. This function was greatly facilitated by the acquisition of a resistant, impermeable covering for the spores, allowing them to travel over considerable distances and to remain dormant for considerable times until encountering conditions favorable to germination and subsequent fertilization.

An important development, in both plants and animals, was separation between the sexes. This occurred very early in the animal line by what is known as *sexual dimorphism,* the division of reproductive functions between two distinct types of mature individuals, males and females, entrusted with the production of spermatozoa and oocytes, respectively, and endowed with appropriate specializations.

In plants, sexual separation was achieved mainly at the spore level. Male spores continued to act in dissemination, mostly in the form of pollen grains, whereas the female spores served to create a static favorable environment in which incoming male spores of the same species would selectively germinate and produce the spermatozoa needed to fertilize the locally produced oocytes. These specializations have reached an extraordinary degree of diversity in the flowering plants. Flowers are veritable traps for catching male spores, developed around the system that produces female gametes and endowed by evolution with myriad features—shapes, colors, scents—that cause our delight but, more relevant to the plants' reproductive success, proved effective in attracting pollinators. Remarkably, most flowers also contain the male reproductive apparatus, but in a form that hinders local fertilization, thus avoiding inbreeding and the attendant perpetuation of the same genes, which is known to be genetically unfavorable.

Seeds and fruits harbor, until germination, the plant embryos issued from fertilized eggs

In all higher plants, the fertilized egg develops into an immature embryo, which soon becomes arrested in its development and enclosed within a resistant casing, together with a reserve of nutrients that are to be used, upon germination, to support the further development of the embryo up to a state where it can exploit environmental resources on its own. Called *seeds,* these structures have a simple covering in the gymnosperms (*gymnos* means naked in Greek), which comprise mostly the pine trees and other conifers. In the angiosperms (from the Greek *aggeion,* covering), or flowering plants, a group that includes the majority of extant plants, the seeds are situated inside fruits, which are formations derived from the flowers and filled, under aspects of astonishing diversity, with rich nutrient stores, which serve for the nutrition of the embryo and have become, thanks to a fruit production that exceeds by far the requirements of reproduction, an abundant and succulent source of food for the animal world (including humans).

Fungi also reproduce by way of spores

Fungi, like plants, rely mostly on spores for their dispersal. The most spectacular manifestation of fungal reproduction is represented by the multifarious mushrooms, the spore-disseminating structures that suddenly shoot up into the open from hidden mycelia that spread their networks below the ground's surface.

In animals, parent mobility favors union between spermatozoa and oocytes

Animals, taking advantage of their motility, developed a great diversity of reproductive strategies. As long as the animals kept to their aquatic birthplace, males often did little more than discharge a swarm of spermatozoa in the vicinity of females, leaving it mostly to the swimming ability of the cells and to their large number to ensure successful encounter with a female's oocytes. Most of the time, females lay their unfertilized oocytes in the same site, so fertilization and subsequent development of the fertilized egg occur in water. Cases are known, however, in which the oocytes are not discharged, but are fertilized and develop inside the female body. Some fish, called viviparous for this reason, produce progeny in this way.

With the passage to land, new mechanisms were required to compensate for the lack of water. The main such mechanism for animals was *copulation,* which is carried out by most land animals, invertebrates as well as vertebrates. The consequence of copulation is that fertilization takes place inside the female body. At first, the ancestral mode of aqueous development prevailed. Impregnated females laid their fertilized eggs in water, where embryological development occurred. Among insects, for example, mosquitoes behave in this way, which explains their predilection for swampy environments. Other insects, as well as many other terrestrial invertebrates, have evolved a great variety of reproductive strategies. Their description is beyond the scope of this book. I shall restrict myself to the land vertebrates, which are of more direct interest to us, as the human species is one of them.

The fertilized egg of vertebrates has always developed in an aqueous medium

Amphibians, which were the first vertebrates to leave water, acted like the mosquitoes, quickly returning to their original medium for embryological development. Frogs offer a familiar example of this behavior. Females, after copulating on land, lay their eggs in water, where the eggs develop into swimming tadpoles adapted to aquatic life. Then, at some stage, signaled by the secretion of thyroid hormone, the tadpoles shed their tail, sprout two pairs of legs, lose their ability to derive oxygen from water, and start breathing air. The mature animals pursue their existence on land, but in the vicinity of water, where, eventually, females return to lay their fertilized eggs. This group of animals continues to thrive in all marshy lands.

The link with water was broken—or rather displaced from external to internal—by the reptiles, thanks to acquisition of a new structure of crucial importance, the *amniotic pouch,* a closed, fluid-filled sac within which fertilized eggs henceforth underwent development; they no longer needed to be laid in a body of water. Usually encased within a hard shell, the eggs could be left on land to continue their development and hatch in the open.

The reptiles bequeathed this reproductive mode to the birds and to the first mammals, the monotremes (such as the platypus), which still lay eggs. A branch then arose, in which egg-laying was replaced by birth at a very early stage of development, which was allowed to continue further within a ventral pouch, or *marsupium,* from which the young could reach the mammary glands to feed. Thus were born the marsupials, such as kangaroos and koalas.

The last major acquisition in this saga was the *placenta,* a remarkable structure that brings in intimate proximity, separated only by the thickness of blood-vessel walls, maternal blood brought in by the mother's circulation and fetal blood conveyed via the umbilical cord, so that nourishment can pass from mother to fetus, and waste products can be unloaded from fetus to mother. Thanks to this development, which is characteristic of most of today's mammals, including humans, development was allowed to continue inside the womb up to a sufficiently advanced stage for the young to be able to pursue an independent existence (with appropriate fostering). Note that, even in this most perfected mode of development, the fetus continues development in its ancestral, aquatic mode within the amniotic pouch. Human birth, as every mother knows, is heralded by the "breaking of the waters."

6
Development

How, in a matter of nine months, does a fertilized egg become the miracle that is a newborn baby? This question has been asked by generations of biologists ever since William Harvey (1578–1657), the English physician who discovered blood circulation, exclaimed, after dissecting a pregnant doe felled in hunting by his patron, King Charles I: "Omnia ex ovo," all (living beings) arise from an egg!

The first accounts of embryological development were purely descriptive

The embryologists who tackled this problem found that the fertilized egg first divides into a small number of almost identical cells, which form a cluster called the *morula,* the diminutive of *morum,* the Latin word for mulberry. These are the widely publicized *stem cells,* called "totipotential" because they can give rise to any cell in the body.

Soon, the morula turns into a hollow sphere, the *blastula,* fitted with a single opening, the blastopore, which went through a remarkable history in the course of evolution. It was mentioned in chapter 3 that the first animals to possess a digestive pouch, such as jellyfish, have only a single opening connecting the pouch to the outside and serving both for the intake of food and for the discharge of waste. Later, the pouch acquired a second opening, turning into a canal, with a mouth at one end and an anus at the other. What was not mentioned in chapter 3 is the remarkable developmental history of these openings. In the first animals with an alimentary canal and in most invertebrates that followed, including all mollusks and arthropods up to the present day, the mouth originates from the blastopore, and the anus from the newly formed opening. At some stage, however, an extraordinary developmental flip-flop initiated a new evolutionary line, with the blastopore henceforth giving rise to the anus, and the new opening becoming the mouth. This conversion from protostomes (mouth first) to deuterostomes (mouth second) initiated events of immense portent. It was followed, in the line where it occurred, by formation of a dorsal structure, or notochord, soon to be replaced by a segmented assemblage of cartilaginous units, the first vertebrae. This is how the vertebrates were born. The significance of this development could hardly be overestimated.

Starting from the blastula, a long succession of highly complex developmental changes occur, following different scripts in different species. In the human species, these steps lead—from embryo to fetus to newborn—to the progressive appearance and shaping of limbs, organs, and other body parts, up to a stage where the newly made organism is ready to leave the protective shelter of the amniotic fluid and sever its lifegiving connection (through umbilical cord and placenta) with the maternal organism and make its entrance into the outside world where oxygen is to be obtained by breathing, food by sucking, and attention by shrieking.

A remarkable aspect of this development, already noted by early observers, is that it goes through stages that recall the evolutionary history of the organism. Thus, the human embryo has gills at one stage, like fish. In the words of the German biologist and philosopher Ernst Haeckel (1834–1919), who was an enthusiastic disciple of Darwin, "ontogeny recapitulates phylogeny," by which he meant that the developmental program of an organism recapitulates the organism's evolutionary history, a view that is no doubt oversimplified but nevertheless proved perceptive.

Experimental embryology began to decipher developmental mechanisms

By the 1950s, this developmental script was known in exquisite detail, not only for humans, but also for chickens, fruit flies, and several other animals. But the script was known mostly in strictly descriptive fashion, with hardly any information on mechanisms, like a movie lacking a soundtrack. This was not for want of trying. In the beginning of the twentieth century, experimental embryologists, led by the German Hans Spemann (1869–1941), had succeeded, by delicate interventions, in identifying "morphogenetic gradients," presumably created by substances, called "organizers," secreted at one site of the embryo and diffusing toward the others. But the nature of these substances, their origin, and their mode of action were totally unknown, until, in the second half of the century, the key to the riddle was revealed almost overnight, at least in principle. The secret turned out to be: *transcription control.*

Development is ruled by transcriptional gene control

We have seen that the instructions stored in DNA—which include, prominently, the directives for embryological development—must be transcribed into RNA in order to be executed, most often by proteins synthesized according to the RNA transcripts. Thus, the process of transcription is the obligatory channel for gene expression. Certain protein molecules, called *transcription factors,* regulate this process. They turn genes on or off, that is, they induce transcription of the genes or block it. Some of these transcription factors even have a graded effect, adjusting the rate of transcription and, thereby setting the extent to which a given gene is expressed. Transcription factors exist in the simplest of bacteria but are enormously more numerous and more important in multicellular organisms. They control the whole of embryological development.

As already seen in chapter 3, all the cells in the body have the same genome. Cells differ, becoming skin cells, nerve cells, liver cells, and so on, by transcription regulation. Turning on certain genes and shutting off others determines the fate of a given cell. This is how identical stem cells differentiate to become the 220-odd different kinds of cells that compose the human body. Thanks to the tools of modern molecular biology, we are beginning to know which genes need to be awakened or silenced in order to make a given type of cell.

This is only a small part of the story, of course. In embryological development, cells do not just differentiate into given types; they become associated in specific patterns, to form tissues and organs, which themselves become organized according to a specific blueprint or body plan. Called morphogenesis, this process is extremely complicated and still poorly understood, as it depends on mysterious signals between and among cells that are only beginning to be unraveled.

Genes are organized by transcription into a hierarchy dominated by master genes

Transcription factors are proteins, which means that they are the translation products of genes, which are themselves subjected to regulation by transcription factors produced by other genes, and so on. There thus exists a hierarchy of genes, which is dominated by "supergenes" that act as master switches for entire developmental programs.

An example is the *eyeless* gene, so named, in the quirky nomenclature devised by geneticists, because it was discovered through a mutation that causes inborn blindness. The gene itself is responsible for the opposite of eyelessness; it controls the complete genetic program of eye formation throughout the animal world. If the *eyeless* gene from fruit flies is injected somewhere in the body of a fruit fly, it induces the formation of a complete, multifaceted fruit fly eye at the site of injection. If injected into a mouse, the same gene will induce the formation of a typical mouse eye at the site of injection. Conversely the mouse *eyeless* gene induces the formation of a mouse eye in a mouse, but of a fruit fly eye in a fruit fly. It is a universal switch. The machinery it sets off depends on the local genome.

Homeotic genes are master genes of central importance

Among the most important master genes are the genes called *homeotic*, which control complex developmental programs that may affect the entire body. These genes share a sequence of 180 bases, called the *homeobox,* which is highly conserved throughout the animal world and even in plants and fungi. This sequence codes for a stretch of sixty amino acids (the building blocks of proteins) whereby the corresponding protein binds to DNA, a prerequisite for its ability to act as a transcription switch.

In primitive animals, there is a single set of homeogenes, which command the development of the body plan. In segmented animals (see chapter 3), there are as many sets of homeogenes as there are segments, aligned along the chromosomal DNA in the order in which the segments follow each other in the body. In simple annelids, such as earthworms, the homeogenes are almost all the same and the segments they control are all identical, except for minor changes in the head and tail. With increasing diversification, each set of homeogenes has evolved on its own to produce increasingly different segments. Like the *eyeless* gene, such homeogenes induce the formation of the kind of segment they code for wherever they end up in the body. This is how investigators working on the fruit fly *Drosophila,* the central object of classical genetic research, have been able, by a single manipulation, to create freaks such as headless, two-tailed flies, animals with an extra pair of legs or wings, or strange monsters sporting legs in front of their heads in place of antennae.

Evolution and development are intimately linked

Note how evolution and development meet in these phenomena. Repetition and differentiation of homeogenes started as an evolutionary phenomenon, which later became inscribed into the developmental program. Often designated by the acronym “evo-devo,” this concatenation of evolution with development was summed up in Haeckel's famous aphorism, "ontogeny recapitulates phylogeny," quoted above.

The discovery of master genes has illuminated a number of evolutionary events of dramatic suddenness. We have already encountered segmentation, one of the most fateful changes in the evolution of animals. Another epoch-making genetic jump, so far unidentified but most likely involving a master gene, is the protostome-deuterostome flip-flop that initiated the line leading to vertebrates (see above).

Incidentally, the history of homeogenes illustrates one of evolution's favorite "tricks": gene *duplication,* which we have seen in the preceding chapter allows one copy of a gene to evolve while the other copy continues exercising its function. The whole history of life is landmarked, from its very beginning, with gene duplications, which lie behind countless evolutionary innovations.

In summary, we are still far from knowing how a fertilized egg produces this miracle that is a newborn baby and maybe never will, considering the awesome complexity of the underlying programs. But, at least, we know the key to the riddle. It lies in the hierarchy of gene-transcription mechanisms.

7
Natural Selection

Charles Darwin (1809–1882), who is often credited for having discovered evolution, was not even born when the transformist hypothesis was first formulated by his grandfather and by Lamarck. What Erasmus Darwin's grandson will forever be remembered for is his proposal that natural selection of hereditary variants is the mechanism by which evolution occurs. Natural selection, contrary to evolution, which is an undisputable fact, may still be viewed as a theory, at least to the extent that it may not be the only mechanism involved in evolution, as we shall see in the next chapter. The actual occurrence and overwhelming importance of natural selection are no longer in doubt.

Unfortunately, evolution and natural selection have become conflated into the single term Darwinism. There is a historical reason for this conflation. The two notions were simultaneously defended in Darwin's major opus, published in 1859 under the ambivalent title *On the Origin of Species by Natural Selection.* In the uproar that followed, Darwin was attacked more for defending evolution, a shocking theory that negated biblical truth and downgraded man to the status of mere animal. His account of natural selection as the basic mechanism of evolution was merely an aggravating circumstance that denied any intervention by God in what was assumed to be a godless process in any case. This confusion has persisted until today, fueling much of the current controversies over evolution. In this and the following chapter, I shall try to clarify the issue.

At the start lies heredity

As a starting point, consider heredity. This phenomenon was known to Darwin and to countless generations before him. Mice beget mice, acorns oak trees, humans babies. The phenomenon goes even further. Children resemble their parents more than they resemble the parents of other children. Today, we know why that is so. Individual blueprints are encoded in DNA, and these blueprints are transmitted from generation to generation by DNA replication. Darwin did not know this. He did not even know the laws of heredity, first formulated in 1866 by an obscure Austrian monk, Gregor Mendel, and appreciated by the scientific community only after Mendel's death in 1884.

Artificial selection exploits the imperfections of heredity for defined purposes

What was known to Darwin, however, and made a strong impression on him, is that heredity is not perfectly faithful; it allows for diversity, a natural circumstance that has been exploited ever since humans started domesticating plants and animals. Look at dogs. The American Kennel Club recognizes 173 breeds. All are dogs; they recognize each other as such, communicate by the mysterious signals particular to their species, and interbreed to give viable offspring. This diversity is human-made. At the start, there was a single variety of wolf or jackal that established some kind of mutually beneficial association with a human group. From then on, breeders created all the existing varieties artificially, using empirically devised methods based on selection of appropriate progenitors. This history, which was repeated with horses, cattle, chickens, cats, and other domestic animals, as well as with a number of plants, goes back to early days of human development prior to any written record.

Today, we know the cause of diversity. It is due to modifications, or mutations, in the DNA, such that a slightly altered blueprint is transmitted from parent to offspring. This, again, was totally unknown to Darwin. But he was keenly aware of the existence of diversity in the living world and of its role in allowing breeders to use artificial selection in an empirically purposeful manner to generate cows that gave more milk, sheep that yielded thicker wool or better meat, horses that ran faster or carried heavier loads, cereals more resistant to cold or drought, and so on.

Malthus introduced the notion of the "struggle for life"

Another major influence on Darwin's thinking was a book by the English clergyman and economist Thomas Robert Malthus (1766–1834), titled *Essay on the Principle of Population as It Affects the Future Improvement of Society.* In this book, first published in 1798, but still popular in Darwin's time, Malthus contrasted the exponential growth of human populations against the linear growth of resources, inevitably leading to a situation where consumers outgrow resources and start competing for them. This situation, embodied in the celebrated phrase "struggle for life," could, according to Malthus, only be avoided by limiting the birthrate, a doctrine that has become known as Malthusianism.

Natural selection lets the "struggle for life" choose passively among the diversity created by the imperfections of heredity

Putting together natural diversity and struggle for life, Darwin arrived at the conclusion, evident for us today but visionary at the time, that, in any natural situation where there is competition for limited resources, those varieties most apt to survive and, especially, produce progeny under the prevailing conditions must automatically become preponderant (fig. 7.1). Hence the subtitle Darwin gave to his book: *The Preservation of Favoured Races in the Struggle for Life.* In this new view, selection, rather than being purposefully directed toward a predefined end, as in its artificial, human-devised form (such as dog breeding and the like), is taken to occur naturally, automatically, without preconceived intention. Here is where Darwin's ideas encountered the strongest resistance, lasting up to the present day; they implied a lack of purpose in nature.

Darwin conceived his theory during his famous voyage, from 1831 to 1836, on the *Beagle,* which notably led him to visit the Galápagos Islands. It is there that he made one of his most perceptive observations, namely that the finches on each island had, during their long geographical isolation, acquired differently shaped beaks adapted, in each case, to the locally available food, a paradigmatic instance of natural selection (fig. 7.2). Nevertheless, Darwin continually put off the publication of his theory, patiently accumulating more evidence that would either support it or disprove it, in the true spirit of objective science. It is only when Darwin was informed that his

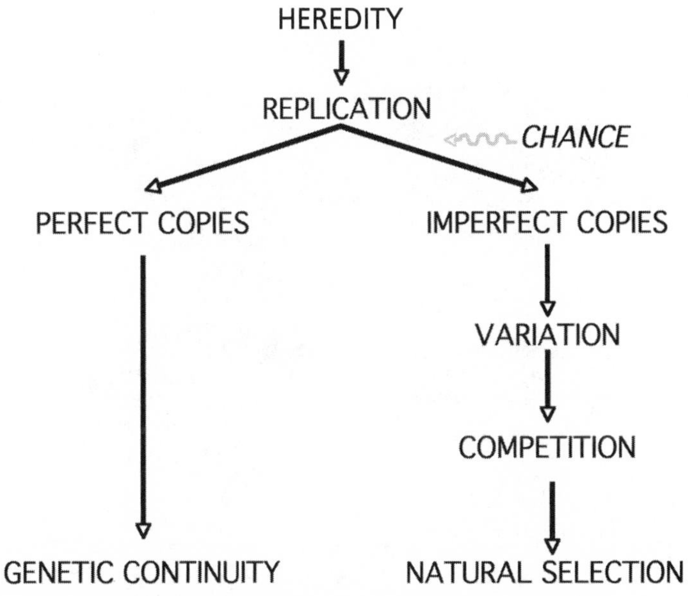

Fig. 7.1. *Natural selection.* This diagram illustrates the obligatory consequences of replication. As long as replication occurs faultlessly, information is faithfully transmitted from generation to generation. It is the source of genetic continuity. When, as must inevitably happen, replication is not faultless, the imperfect copies produced generate diversity, leading to competition among the variant forms for available resources. As a result of this competition, the variant form or forms best able to survive and, especially, produce progeny under prevailing conditions necessarily emerge. This is natural selection. A fact of capital importance is that copying imperfections (mutations) are chance events or, more precisely, occur in a manner in no way related to any sort of adaptive anticipation of a future environmental challenge (see intelligent design, chapter 8).

colleague Alfred Russel Wallace (1823–1913) had conceived the same theory that he decided to rush into print. Much has been written about the poor place accorded to Wallace by posterity. He certainly deserves credit for having independently thought of natural selection as an evolutionary mechanism. But to Darwin

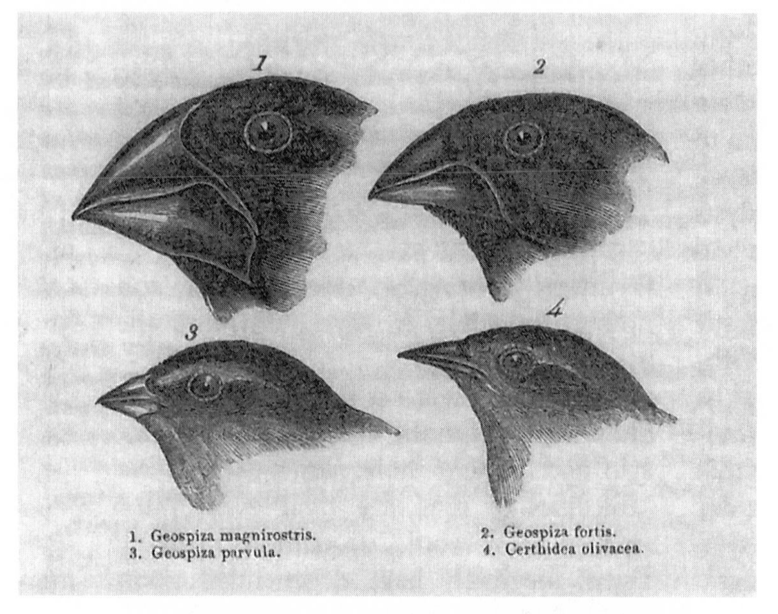

Fig. 7.2. *Natural selection illustrated.* The Galápagos finches. One of the main findings made by Darwin during his historic voyage on the *Beagle* (1831–1836) was that the finches on each of the Gálapagos Islands had different beak shapes, adapted to the available food. He reasoned that these adaptations had emerged by natural selection during the long period of isolation of the birds. From Charles Darwin's *On the Origin of Species.*

goes the immense merit of actually giving body to the idea by laboriously building a monumental set of evidence in its favor.

Natural selection acts under our very eyes

Today, the reality of natural selection leaves no room for doubt. Among countless examples arrayed in its support one can cite, as having taken place under our very eyes, the many cases of antibiotic resistance that developed in only a few decades after penicillin was first used during the last war, opening the way to the discovery of a string of other, similar drugs. In each case, introduction of a new antibiotic for therapeutic use was rapidly followed by the development of pathogens resistant to the antibiotic, clearly originating from rare, naturally resistant varieties that prevailed when exposed to doses of the antibiotic that killed the more sensitive bacteria. Paradoxically, hospitals have become sites where some of the most dangerous infections may be caught, as they provide an environment particularly enriched in antibiotics and, therefore, particularly conducive to the selection of the most antibiotic-resistant pathogen varieties. Similar cases of resistance have been observed with herbicides, insecticides, and other pesticides, even with chemotherapeutic agents, to which resistance may build in a matter of months in the cancer cells of treated patients.

Another often-quoted example of natural selection is industrial melanism, a phenomenon that affected some English peppered moths (*Biston betularia*) that exist in two different varieties, one white, the other black. In the nineteenth century, when smoke and soot produced by the Industrial Revolution covered all surfaces with a black coating, the white moths almost completely disappeared, while the black ones flourished. The situation has reversed since the end of the Second World War, when laws were enacted to clean the air. The explanation is simple. Predators that feed on moths more readily detect the white ones on a dark background and the black ones on a light background.

The mutations subjected to natural selection are accidental events devoid of finality

A key trait of natural selection, already suspected by Darwin and now confirmed by all that has been learned since, is that the mutations on which natural selection operates are due to *chance* (see fig. 7.3). Thus, one cause of mutations, inevitable, as well as unpredictable, is faulty replication. The molecular mechanism that drives this process is of astonishing fidelity, one wrongly inserted base in about one billion, the equivalent of copying the *Concise Oxford Dictionary* fifty times by hand, making a single mistake! Yet, replication mistakes are a significant cause of mutations because of the huge number of bases contained in genomes. Thus, every time a human cell divides, about half a dozen errors are made in the replicated genome. Fortunately, most of these mistakes are of no consequence. But they contribute to genomic diversity through the germ line.

Other sources of mutations are chemical alterations of DNA molecules caused by physical agents, such as ultraviolet light, X-rays, or radioactivity, by chemical agents, called mutagenic for this reason, by biological agents, such as viruses, or by faulty rearrangements in the course of recombination. Most of these agents are also carcinogenic; they cause cancer, which is often due to a mutation in the cell that initiates formation of the tumor. All these modifications are due to specific causes; but they are accidental, in the sense that they are not intentional. They are not directed toward a goal, which would be, for instance, adaptation to certain outside circumstances or accomplishment of a given evolutionary step. We shall see in the next chapter that this is a crucial point with respect to the theory of intelligent design. Note, however, that natural selection has allowed emergence of pseudo-intentional mechanisms whereby, for example, a stress situation increases the frequency of mutations, thereby enhancing the possibility that a mutation will occur that produces progeny able to survive the stress.

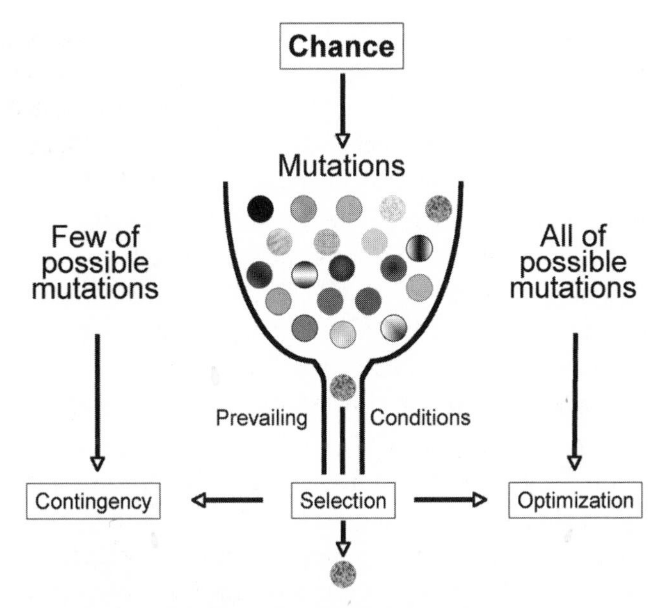

Fig. 7.3. *The evolutionary lottery.* A schematic representation of natural selection (see fig. 7.1), illustrating the two possible extreme outcomes, depending on whether chance offers only a small subset or an essentially complete array of all possible mutations to screening by natural selection. The phenomenon is ruled by contingency in the first instance, by optimization in the second.

The role of chance in evolution is limited by stringent constraints

Given that chance offers natural selection the array of mutations on which it will operate, important implications follow (fig. 7.3). One self-evident implication is that only a mutation included in the array offered by chance can be selected. There could be better solutions to the environmental challenge to which the organism is exposed, but if chance does not provide the appropriate mutations, none of those solutions can materialize.

An important corollary of this implication is that the probability of a response being the best possible one depends on how many mutations are offered by chance. If all possible mutations are offered, then the response will be optimal and, therefore, reproducible if the same challenge were to arise again. This, at first sight, would seem to be an extremely unlikely situation. Somehow, when it comes to mutations affecting genomes of up to billions of bases, we intuitively think of an immense number of possibilities, of which only a small subset actually takes place at any given time. This has long been the received truth among leading evolutionists, who have all insisted on the utter contingency of the evolutionary process. The late American paleontologist and best-selling author Stephen Jay Gould vividly illustrated this view in his famous tape analogy: rewind the tape and allow it to be played again, and a completely different story will unfold.

This view ignores the enormous numbers of individuals and generations that may participate in evolution and the very long times involved. Just to give an example, a simple calculation shows that it takes twenty billion cell divisions for a given base in a given site of a genome to be replaced with a 99.9 percent probability by another given base (point mutation) as a result of a replication mistake. This may seem like a huge number. Actually, it corresponds to the number of divisions that take place in two hours in our bone marrow in the course of red blood cell renewal. In science, as in other human endeavors, intuition may be misleading.

Also, the structure of genomes is often such as to limit the number of mutations. Remember the example given above of mutability increasing in a stress situation. This could not be so if genomes were liable to suffer all possible mutations all the time. Another factor that warrants consideration is that many different mutations often produce the same effect.

Because of these reasons, favorable mutations happen more frequently than one might be inclined to expect. In the evolutionary game, tempting chance may actually pay. Experimenters have long been empirically aware of this, when trying indiscriminate exposure to a mutagen to achieve a given result—and often succeeding. Thus, in the days when penicillin was first produced, a major breakthrough was achieved by massively exposing samples of the penicillin-producing mold to X-rays, "carpet-bombing" fashion. Mutants producing twenty times more penicillin than the parent variety were obtained by this simple device, thus suddenly making the miracle-drug available at affordable cost. Many other such examples are known.

Cases of optimizing selection are more frequent than long believed

When we look at the facts, we do indeed find that there must have been many cases in which chance has offered natural selection a large enough sample of the possible mutations to allow a near-optimal outcome. Look at biodiversity, this extraordinary collection of millions of different living species found in almost every possible habitat, from rain forests and swamps to the driest of deserts, from boiling volcanic springs to polar ice fields, from pristine mountain streams to drying brine, stinging acids, caustic alkalis, or heavy metal–laden industrial wastes. This diversity is often cited as evidence for the contingency of the evolutionary process. What it illustrates much more eloquently is the incredible adaptability of living forms, which, in the face of innumerable different challenges have so often succeeded in coming up with a survivable response.

Particularly impressive in this respect are the many instances of mimicry, the phenomenon by which some animals have acquired appearances that make them almost indistinguishable from their environment, for example, insects that look like leaves or branches, fish that look like the bottom of the sea, and so on. Obviously useful as a protection against predators, these phenomena are astonishing, as they imply a multistage evolutionary process—a beetle cannot suddenly come to look like a leaf—in which, at every stage, the organism underwent a small change that made it look a little more like its environment, enough to give it a selective edge over its unmutated congeners. What is remarkable is that a favorable mutation took place at each of those stages.

Also remarkable are the many instances of animals that have separately developed the same specializations in the face of the same challenges. Anteaters, moles, felines dependent on hunting, and herbivores built for speed look very much the same in widely different parts of the world, where they evolved in completely independent fashion. This phenomenon, called "evolutionary convergence" or "convergent evolution," has attracted considerable interest in recent years. It is cited in recent books by Simon Conway Morris (*Life's Solution: Inevitable Humans in a Lonely Universe,* 2003) and Richard Dawkins (*The Ancestor's Tale: A Pilgrimage to the Dawn of Life,* 2004), two British evolutionists of the younger school, as convincing proof of the inevitability and repeatability of many evolutionary happenings. Dawkins even goes so far as directly to counter Gould by stating that if the tape were replayed, essentially the same story would unfold. Interestingly, Conway Morris defends a religious vision, whereas Dawkins is a militant atheist. So, it is not ideological bias that prompts either of these scientists to assert this view.

Evolution is largely molded by environmental conditions

One last feature of natural selection that deserves emphasis is its dependence on environmental conditions. Whatever genetically determined attribute is naturally selected and therefore survives is directly related to the kind of challenge the evolving organism is exposed to. Thus, if increasing temperature is the challenge, the most heat-tolerant form present among the offered varieties will be selected. In the face of increasing cold, on the other hand, the form best adapted to a low temperature will emerge. And so on. In all the examples of optimization given above, whether biodiversity, mimicry, or convergence, the common determinant has been adaptation to a given environment. If evolving life had encountered different environmental conditions, it no doubt would have generated a different biosphere. Here, indeed, the word "contingency" finds its deserved place. The history of life on Earth is undoubtedly unique, even if there should be billions of life-bearing planets, as no planet can go through precisely the same kind of history as another one, given all the erratic cosmic, geological, climatic, and biological upheavals it would undergo.

Certain evolutionary events could be potentially present in genomes and made manifest by favorable environmental conditions

Such being the case, while the history of each planet is certainly unique, the events that compose it are less so. Tectonic movements, volcanic eruptions, dryness, floods, glaciary periods, tropical episodes, even the fall of meteorites are events that have a considerable likelihood of happening at one or another moment in the history of a planet situated, like Earth, in the habitable zone surrounding its sun.

Within such a context, one may wonder whether certain key events of the history of life had not become almost necessary at the stage reached by evolution and waited, so to speak, for the environment to provide them with an opportunity to take place. As an example, if the fall of a meteorite had not precipitated the disappearance of dinosaurs sixty-five million years ago, is it not conceivable that those monstrous animals were in any case slated to disappear, together with the luxuriant vegetations from which they drew their subsistance, and that, if not the fall of a meteorite, some other accident would at some time have triggered their eradication. As we shall see, the same kind of question arises with respect to the advent of humanity.

Thus, the notion of pure contingency as the driving force of evolution should perhaps be replaced by that, more subtle, of coincidence between an evolutionary stage potentially capable of leading to a critical step in the history of life and the environmental conditions needed for this step to be accomplished. In such a case, the main lines of the evolutionary course would be more or less probable depending on the likelihood of such coincidences.

Perhaps the picture, long taken for granted by a majority of evolutionists, of an evolutionary process largely dominated by the vagaries of the environment should be replaced by that of a process dominated, at least in its main lines, by its internal dynamics but dependent on the environment for the actualization of its potentialities. We have seen that several modern evolutionists lean in favor of such a conception.

8
Other Evolutionary Mechanisms

Darwinian selection constitutes the main mechanism of biological evolution. As we saw in the preceding chapter, the evidence supporting this statement is overwhelming. But natural selection is not solely in charge. Several other mechanisms have been proposed that, while not replacing natural selection, may play a significant additional role.

Lamarck advocated the heredity of acquired characters

The first evolutionary theory to be conceived antedates natural selection by half a century. It was elaborated, together with evolution itself, by the Frenchman Lamarck, in his 1809 opus *Philosophie zoologique.* This theory postulated that useful traits acquired during life could be transmitted to progeny. A favorite example cited by Lamarck is the giraffe, which he saw as having acquired its long neck as a result of efforts by generations of giraffes to reach the highest branches of trees. A

Darwinian explanation of the same fact would be, for example, that giraffes that happened by chance to be born with hereditarily transmissible longer necks had a better chance of surviving and transferring this property to their offspring because they alone could still feed when the lower branches, which were the only ones that giraffes with shorter necks could reach, had no leaves left.

Lamarck's theory of inheritance of acquired characters lost a lot of ground—though not without a struggle—after natural selection was proposed by Darwin. In the beginning of the twentieth century, the Austrian biologist Paul Kammerer announced that he had succeeded in causing toads that copulate on land to acquire the "nuptial pads" whereby the males that copulate in water grip the slippery body of the female. He claimed to have accomplished this transformation simply by forcing the animals to copulate in water, generation after generation. Accused of falsifying his results, Kammerer eventually committed suicide. This did not prevent the British writer of Hungarian origin Arthur Koestler from defending Kammerer's memory and presenting him as a victim, in *The Midwife Toad,* published as late as 1971, well after the advances of molecular biology had invalidated the foundations of Lamarckism.

A much more dramatic instance of Lamarckian fraud was perpetrated under Stalin by the Soviet agricultural biologist Trofim Lyssenko, who claimed to have converted winter wheat into the more productive and faster-growing spring wheat by the simple device of "vernalization," which depends on soaking and chilling the grains. When this claim was contested by geneticists, Lyssenko used the apparent agreement between his theory and Marxism as an argument to have the dissenters condemned, with catastrophic consequences for the future of genetics and of agriculture in his country. This scandalous affair caused a number of prominent leftist scientists in the West to cut ties with Soviet Russia and resign from the Communist Party.

DNA cannot be a vector of Lamarckian heredity

The inheritance of characters acquired by the parents' experience was ruled out as a possible evolutionary explanation by the findings of modern molecular biology, at least as concerns transmission via DNA. There is no way whereby such an acquired trait could travel up to DNA and become hereditarily encoded in it. The pathway from gene to trait is strictly one-way. Crick has gone so far as to call this rule the "Central Dogma." This choice was unfortunate, since science knows no dogmas, but it underlines the strict incontrovertibility of the rule. The Darwinian mechanism, on the other hand, is perfectly compatible with modern biology and supported by it.

Lamarck still has a few rearguard defenders, especially in France, where he has become something of a national symbol in the old struggle against Anglo-Saxon cultural hegemony. Not so long ago, after giving a lecture in a French university town, I was treated to a lengthy harangue by an elderly biology professor, extolling the qualities of Lamarck and lamenting his abandonment, which the professor saw as a betrayal in favor of the Englishman Darwin.

Cases of Lamarckian heredity that do not involve DNA exist

Note, however, that the ban against Lamarckism concerns only DNA-mediated heredity. Recent years have seen discovery of several other forms of heredity susceptible to a Lamarckian explanation. Biological membranes offer a revealing example. Cellular membranes grow by accretion, that is, incorporation of new components into preexisting membranes. In this process, the pattern of the recipient membrane influences the "choice" of the new constituent to be inserted. A membrane modified by usage could thus induce the insertion of a different constituent and change the pattern of the new material added to it. This material would, by the same mechanism, transmit this modification to progeny, with as consequence the inheritance of an acquired character, in Lamarckian fashion. Another, similar, example concerns the pattern of implantation of swimming appendages, called cilia, on the surface of some protists named paramecia. Let this pattern be changed by environmental influences, and later progeny will display the new pattern.

The discovery of prions reveals a particularly dramatic case of shape transmission. Prions are infectious agents of protein nature that owe their pathogenicity to their abnormal shape. Their molecular sequence is normal, but the manner in which the chains are folded is abnormal and—accounting for the infectious nature of the molecules—can be transmitted to normal molecules by contact. Mad cow disease and its human version, Creutzfeldt-Jakob disease, which made headlines a few years ago, are typical instances of prion diseases. The importance of this kind of transmission in normal heredity is not known but could be significant. Future work will tell.

Yet another instance of possible Lamarckian inheritance is represented by a set of phenomena recently grouped under the term "epigenetics," a term long used in an entirely different meaning by developmental biologists and neurobiologists (see chapter 16). In its new meaning, epigenetics refers to a number of inheritable traits that are not written into the DNA sequences but accompany the DNA in germ cells and influence subsequent events in the fertilized egg. Such traits include the blockage of certain bases by chemical groups (for example, methyl groups) or the manner in which DNA is combined with the local proteins, or histones, in the chromosomes. Some of the most exciting new findings are being made in this area.

Genetic drift accompanies evolution without selection

Another non-Darwinian form of inheritance concerns traits that are transmitted without being selected. Many genetic mutations seem close to neutral, with little selective value, whether positive or negative, and just accompany the others by inertia, so to speak, because of the simple fact that their elimination is not sufficiently advantageous. The Japanese theoretician Motoo Kimura has developed a mathematical theory, under the name of "genetic drift," that explains how mutated genes are inherited in this way. Many drifting genes exist. The best proof of this is given by the genetic diversity exploited by molecular phylogenetics. We have seen (chapter 1) that this technology uses the sequence differences among genes that play the same role in different organisms. The very fact that such differences exist, and actually do so on a large scale, is proof that many different versions of the same gene can exist and, apparently, perform satisfactorily. Whether these mutations are truly neutral is, however, debatable. For example, modern medicine has identified a number of human genetic variants that affect the probability of falling victim to a disease, such as diabetes or breast cancer. Such "risk genes" are not neutral; they may significantly influence the life span of the individuals concerned. On the other hand, to the extent that the genes do not affect fertility—the diseases they influence often break out late, after the individual has ceased to be reproductively active—the genes are indeed neutral with respect to natural selection.

Self-organization could theoretically drive evolutionary events

An evolutionary factor, unrelated to heredity and believed by some of its proponents to be as important as natural selection, lies in the ability, attributed to certain living systems, to spontaneously settle from an initially unstable situation into a stable, organized pattern. Variously referred to as "self-organization," "autopoiesis," "order out of disorder," and the like, the phenomena underlying this ability have been the object of many ingenious theoretical studies but little experimental work so far.

Were some key evolutionary steps guided by "intelligent design"?

One last alternative to natural selection has been proposed under the name of *intelligent design*. It warrants attention, not because it offers a valid scientific explanation of evolution, which it does not, but because of the political and educational issues it has generated, especially in the United States but also, in recent years, in France and other countries. Intelligent design should not be confused with creationism. Theoretically, intelligent design presumes no biblical roots. Many of its defenders accept evolution. Some even accept natural selection. All they claim is that natural selection does not account for everything and that certain evolutionary steps cannot be explained naturally and must have required supernatural intervention.

This view is not new. It was known in the past as finalism, or teleology, a doctrine close to vitalism that sees life as a goal-directed process (which implies someone or something that does the directing). This is, indeed, the appearance life gives to the observer and is reflected in the common language: we say that the lungs are *for* breathing, the stomach and intestine *for* digestion, and so on. Replacing such statements by "the lungs, or the stomach, happen to be such that they can function in breathing, or in digestion" is counterintuitive but more consonant with current thought, which sees these organs as emerging by natural selection because they were accomplishing functions useful to survival and multiplication of the organisms in which they first appeared. The term "teleonomy" is often used, in opposition to "teleology," to qualify the *apparent* purposefulness of biological structures.

Defenders of intelligent design use several arguments. One, by the American biochemist Michael Behe, is "irreducible complexity," the property of systems made of several parts, each of which needs specific features to fit within the whole. He gives the "simple mousetrap" as a model of irreducible complexity and goes on to cite complex biochemical systems, such as blood coagulation or motor organelles like cilia and flagella, as examples of assemblages that could not have come together without the assistance of some entity that conceived the machinery, foresaw its use, and designed the various parts accordingly. Two centuries earlier, the English theologian William Paley made a similar argument in his famous watchmaker analogy, which served as basis for his *Natural Theology; or, Evidences of the Existence and Attributes of the*

Deity Collected from the Appearances of Nature, which was first published in 1802: you find a watch and deduce that there must have been a watchmaker.

The New Zealander Michael Denton cites as evidence of intelligent design certain evolutionary transitions, such as the passage from reptile to bird. He points out that so many changes had to take place concurrently, in skin, bones, lungs, and other organs, that they could have happened only under the instructions of a designer who knew what the final product would look like.

William Dembski, an American mathematician, evokes the familiar sequence paradox. The calculation is simple. Take twenty different kinds of letters (representing the twenty kinds of amino acids with which proteins are constructed) and use them in all possible combinations to make words. The number of different possible words, known as the "sequence space," increases with the length of the words. It is four hundred for two-letter words, eight thousand for three-letter words, in general, 20^n for words of *n* letters. Many protein "words" have lengths of three hundred or more amino acid "letters," which corresponds to a minimum of 20^{300}, or 10^{390} different words. There is simply no way such a number could be represented, except by filling an entire paragraph with the figure 1 followed by 390 zeros, which is essentially meaningless. It dwarfs the biggest natural numbers we can think of, such as the number of stars in the universe (10^{22}) or even the number of elementary particles in the universe (10^{42}). Even the number of different fifty-letter words (20^{50}, or 10^{65}) already exceeds this range. For Dembski—and for a number of theoreticians who have made the same calculation before him—the conclusion is clear: proteins could not, without guidance, have reached the "infinitesimally minuscule" spot they occupy in their "immeasurably immense" sequence space.

At first sight, such reasonings may be appealing and even seem compelling. Take a case such as mimicry (chapter 7). To an observer faced with an insect that can hardly be distinguished from the leaf on which it sits, the simplest explanation that comes to mind is that of an agency that modeled the insect as a copy of the leaf. The natural explanation, which posits a very large number of small steps in which the ancestral insect each time became a little more leaflike, enough so to gain a selective edge over the others, is, in a way, much harder to swallow, especially if one adheres to the notion of a creator. A God who created life and endowed it with its properties should know enough about genetics and its applications to be able to carry out the changes that make a beetle look like a leaf and help it avoid predators.

It is thus no wonder that intelligent design has been so warmly received among various religious circles, including those that reject biblical creationism. The Catholic Church, for example, has recently been leaning in this direction, first in the words of the Austrian cardinal Schönborn, later endorsed by the pope himself. This is understandable and even logically consistent. For one who believes in a God capable of granting prayers and even performing miracles, directly or through persons of exceptional saintliness, it is not difficult to imagine this God occasionally giving evolution a nudge in a direction of his choice, especially if, as is firmly asserted by Catholic doctrine and several other theistic religions, one believes this choice to include the advent of humankind. But that is not the question. It is, even for the believer, to know whether the nudge was necessary or whether events could have happened naturally. We shall see that the advances of sciences do not favor intelligent design.

In the United States, the issue has become political, even legalistic. The question is not so much whether intelligent design is right or wrong, but whether its teaching in schools as an alternative to natural selection contravenes the separation between church and state. From the didactic point of view, this way of handling the situation ignores the fact that intelligent design is advocated by a significant number of people and the object of much public debate. My preference is to allow discussion of intelligent design and thereby demonstrate why it fails and why Darwinian natural selection succeeds in showing how evolution works. But such discussions would perhaps put an unfair burden on teachers and would probably be even more controversial than banning the subject outright from the curriculum.

Refuting intelligent design is easy and does not require any specialized knowledge. Intelligent design is simply *not a scientific theory.* Science is based on the working hypothesis that things are naturally explainable. This may or may not be true. But the only way to find out is to make every possible effort to explain things naturally. Only if one fails—assuming failure can ever be definitely established—would one be entitled to state that what one is studying is not naturally explainable. The entire history of science, including the dramatic advances of biology in the last fifty years, is there to validate and consolidate the naturalistic postulate. Now is hardly the right time to abandon this cornerstone of the scientific endeavor. Yet, this is exactly what intelligent design does, by stating a priori that certain evolutionary events are not naturally explainable, thus ruling out the possibility of ever finding the explanation if there is one (see also chapter 2). Just tell this to the students and let them draw their own conclusions. It has nothing to do with religion.

At the scientific level itself, the arguments put forward in support of intelligent design can easily be dissected, as many critics have done, and shown to rest on oversimplified views of the evolutionary process, which ignore the immense times taken, the circuitous pathways followed, and the large numbers of individuals and generations involved. Complex biochemical systems, for example, have arisen from simpler ones by way of many steps that are only now beginning to be identified. Similarly, evolutionary pathways that have often been viewed as almost miraculous are now being elucidated. The history of the eye, for example, which has evoked so much wonder and puzzlement, has been largely reconstructed, starting from a small light-sensitive spot in some primitive organism and branching out into at least six different directions.

As to the sequence paradox, we have seen in chapter 4 that the structures of present-day proteins and other observations indicate that these molecules started as short chains. It is conceivable that the molecules were short enough for all possible sequences, or almost, to be realized (by way of their genes) and submitted to natural selection. Combination of the sequences retained by this first screening could have yielded an essentially complete set of the longer sequences achievable with the starting sequences, leading once again to pruning by natural selection, and so forth, up to the sequences of hundreds of amino acids that prevail today. In other words, proteins would not have reached in one shot the "infinitesimally minuscule" spot they occupy in their "immeasurably immense" sequence space, as assumed by Dembski, but by a stepwise pathway subject at each stage to natural selection.

Much remains to be discovered, but so much has already been found that one can only be urged to look for more, rather than give up and invoke supernatural influences of unknown nature.

III
The Human Adventure

Introduction

We enter the last episode of the history of life, the one that concerns us most directly, as we we are its outcome (most likely provisional, as evolution is far from finished). In this part of the book, I briefly retrace the main steps of hominization, paying special attention to the development of the brain, which is its most outstanding event, and to the latest developments that have led to today's world, totally dominated—and threatened—by the inordinate success of the human species.

I conduct this analysis in the light of the notions that are available on the earlier history of life and on its mechanisms, with as main source of illumination the "beacon" of natural selection. I thus arrive at the book's central theme: "original sin" reinterpreted in the light of knowledge, namely the genetic flaw imprinted into human nature by natural selection.

9
The Emergence of Humans

Let us go back about three and a half million years. On the human scale, this is an immense time span: thirty-five hundred millennia, five hundred times the duration of the whole of recorded human history. But, in the framework of the history of life (more than three and a half million millennia), or even of animal life (six hundred thousand millennia), it is little more than a brief coda.

Africa is the cradle of humankind

The site of our flashback is in Africa, in the arid Laetoli region in northern Tanzania. The focus of our attention is a small band of strange creatures of mixed ape-human appearance, revealing their ape kinship by their short size (about four feet), dark, hairy skin, flattened forehead, projecting jaws, and small cranial volume, but walking upright on two legs and holding things in their hands, like humans. We know of them through an amazing set of footprints almost miraculously preserved in solidified volcanic ash, discovered in the late 1970s by the renowned South African couple Louis and Mary Leakey. We have some idea of what these creatures looked like from the world-famous Lucy, the nearly complete skeleton of a young female unearthed in 1974, in the Afar region of Ethiopia, two thousand miles to the north of Laetoli, by a team including the Frenchman Yves Coppens and headed by the American Donald Johanson, who named it after a Beatles song that was incessantly played in their camp, "Lucy in the Sky with Diamonds." Lucy and her congeners are known under the misleading name of *Australopithecus* (literally "southern ape"), which was invented in 1925 by the Australian anthropologist Raymond Dart, who chose it, not because of his nationality, but because South Africa is where he discovered the so-called Taung child, which is the very first prehuman fossil found in Africa, the first clue pointing to this continent as the cradle of humanity.

Where did *Australopithecus*—or rather, australopithecines, as there were several kinds—come from? To answer this question, we must move back several more million years, to the time when the African continent had just been split from north to south by a deep tectonic fracture that gave rise to the Great Rift Valley. This upheaval dramatically altered the local landscape and climate, especially east of the cleft, where the forest gave way to savannahs and to vast, barren areas, episodically exposed to rainfalls that transiently filled the dry riverbeds with rushing streams. This change in the landscape played a decisive role in the advent of humankind, according to a theory, popularized under the name of "East Side Story," proposed by the French anthropologist Yves Coppens (encountered above as a member of the team that discovered Lucy) and endorsed by a number of investigators.

According to this theory, about seven million years ago, a small band of apes, originally belonging to a forest-dwelling group ancestral to present-day chimpanzees, found themselves cut off from the bulk of the group and trapped east of the Rift, where they were exposed to challenging new environmental conditions. They responded to these challenges—so the story goes—by the acquisition of bipedalism, a mode of locomotion that helped them walk economically through the savannah, raise their eyes above the tall grass to spot dangerous predators, and use their hands to carry objects and fashion tools. Bigger brains, better able to coordinate physical activities and to design survival strategies, also evolved over time and helped these early hominids to prevail in their new environment.

This attractive theory has recently received a serious blow from another Frenchman, Michel Brunet, who has discovered what look like prehominid remains—mostly cranium pieces showing some human features—dating back as much as six to seven million years, in the Djurab desert of the Chad region, far to the west of the Great Rift Valley. Called *Sahelanthropus tchadensis* by its discoverer, to stress its human character, and nicknamed "Toumaï," a name meaning "life hope" given locally to children, the new fossil has obviously undermined the "East Side Story" scenario. But its significance should not be overestimated.

We must remember that the prehistory of our species is known to us only through an extremely small number of fossil fragments—often no more than a few teeth, a jawbone, or a femur—tiny bits of evidence left over huge expanses of time. It is like having a kneecap or wisdom tooth from Moses as sole trace of the whole of recorded human history. Anthropologists have proved remarkably perspicacious detectives, capable of squeezing information out of even the slightest of clues, such as a scratch on a tooth or the tip of a toe bone. Nevertheless, linking these sparse fragments of evidence into a coherent historical narrative is still a highly risky exercise. Considering the frequency of evolutionary branching, on one hand, and of evolutionary convergence, on the other, to mention only the most obvious causes of uncertainty, there is little assurance that the clues refer to a single story. It is certainly impressive, in this respect, to find that all the most ancient pre-human vestiges discovered so far, with the exception of Toumaï, have been found in East and South Africa. Until proven otherwise, this part of the world appears as the birthplace of our species.

Note that the African origin of humankind is hardly in doubt. Not a single sign of hominid presence earlier than two million years ago has been detected anywhere else in the world. The question is whether the adventure unfolded only in the eastern part of the African continent. Focus on this area could well be due to chance factors, such as fossil preservation and accessibility, the influence of earlier findings on subsequent research, the whim of investigators, their origin and nationality, the political constraints on their work, not to mention an accidental observation by a completely naïve person, as has happened several times (in Lascaux, for example). Other regions, such as Chad, could well hold some surprises in store for future investigators.

They were not yet human, but they already made stone tools

The second stage in this extraordinary saga, still restricted exclusively to Africa, took place around two and a half million years ago, with the appearance of more humanlike forms called *Paranthropus* ("next to human") or, anticipating the third stage, *Homo*, which is Latin for human. Characteristic of this stage is a sizable expansion of the brain volume, accompanied, presumably, by enhanced manual skill and the related development of stone tools.

Tools, contrary to what one might suspect, are not a

human invention. Many animals use stones or sticks to break, dig, or even kill. The most intelligent primates, such as chimpanzees and bonobos, are even known to fashion a tool with what seems like a purpose in mind. A typical example, first reported by Jane Goodall, the pioneer of primate research in the wild, is that of chimpanzees denuding a branch, sticking it into a termite nest, waiting a few moments, and pulling it out again to eat the insects that cling to it. It is difficult not to see some intentionality behind such behavior. The devising of tools by Stone Age hominids is no more than a continuation and elaboration of this primate faculty, but manifesting considerably greater foresight and skill.

These tools can be seen in any museum of natural history or anthropology. You will find them ranged in orderly fashion, from roughly hewn pebbles, with only a few flakes chopped off, to exquisitely honed axes and blades of various shapes, obviously adapted to specific uses. What you may not realize is that the evolution of tool-making skills covers more than two million years, more than two thousand millennia. In comparison, it took us only a small fraction of a millennium to move, for example, from the abacus to the slide rule, and even less from the slide rule to the computer. The difference is clear. The jump from abacus to computer was accomplished by the same kind of humans with the same kind of brains, who shared and transmitted their experience by communication. Computers are products of *cultural* evolution. Prehistoric stone tools are the products of *biological* evolution. The first tools were made by prehumans, appropriately called *Homo habilis,* handy man, with brains only half the size of our brains today. The improvement of the tools reflects the increase in the size and complexity of the prehuman brain and the gain in manual dexterity that went together with this increase by way of back-and-forth interactions between brain and hands. Cultural transmission, of course, was also involved. Even chimpanzees have a culture. The point is that mental and manual ability were pace-limiting factors of technological progress. Hence the extreme slowness of tool development.

Prehumans started out of Africa for the first time some two million years ago

The next stage on the road to humankind was inaugurated some two million years ago by two new groups, *Homo ergaster* (from the Greek *ergon,* work) and, especially, the anachronistically named *Homo erectus* (his ancestors were upright for more than two million years), who persisted for a particularly long time. These newcomers could almost be called human. They were tall (about six feet), had brains up to two-thirds the size of a modern human brain, probably had lost most of their body hair, walked and even ran, if necessary, on two legs, and manufactured sophisticated tools. They also formed more closely knit and socially organized bands than did their predecessors, built shelters, centralized tool-making and butchering in special areas, and, at some stage, learned to use the fire ignited by lightning, husband that fire, and, eventually learn to light it on their own. They no longer lived only on gathered fruits, roots and tubers, and on carrion stolen from predators; they started hunting on their own. Most impressive, they began to move out of Africa.

Those migrations are amazing, considering the physical fragility of the migrants. Brain power, no doubt, was a major asset in their success. Bands of *Homo,* presumably driven by climatic changes and by the movements of the herds from which they drew their subsistence, spread out east of North Africa, to invade much of Asia, up to China and Indonesia, where they left some of the first hominid fossils to be discovered outside Europe, including the famous *Java Pithecanthropus* (literally ape man), so named by the Dutchman Eugene Dubois, who discovered it in 1891, as well as *Sinanthropus,* or Peking man, which contributed to the celebrity of the French Jesuit Pierre Teilhard de Chardin.

Other bands moved north and northwest to create settlements in many parts of Europe. A late wave is represented by *Homo heidelbergensis*—first discovered in Germany, as its name indicates, but present over much of Europe, Africa, and even Asia—which was still around less than half a million years ago.

A second wave of migrations started once again out of Africa

Surprisingly, these first invading groups, which at some time occupied much of the Eurasian continent, all died out, leaving Africa as the sole point of departure of the final stage that gave rise to modern humans. This question has long been disputed, pitting the partisans of polygenism, who believe that several distinct lines have independently produced representatives of *Homo sapiens* in different parts of the world, against those who defend monogenism. As long as only fossils were available to settle the issue, the matter remained debatable, though with majority opinion slowly veering toward monogenism, much to the satisfaction of those who put their faith in the Bible. The clinching argument has come from molecular phylogenies.

We have seen (chapter 3) that mitochondria, which are the major centers of oxidative energy production in the vast majority of eukaryotic cells (including human cells), are derived from bacteria that were adopted as endosymbionts. The most convincing proof of this origin is provided by the presence in mitochondria of small amounts of DNA, inherited from the ancestral bacterium. We have also seen (chapter 5) that, in sexually reproducing organisms, the male sperm cells lose their mitochondria in the course of maturation, so that these organelles are transmitted from generation to generation exclusively through the female line, by way of oocytes. Comparative sequencing of samples of human mitochondrial DNA collected in various parts of the world has allowed a reconstruction of this history, which leads back to a single ancestral female—the so-called mitochondrial Eve—who lived somewhere in Africa about two hundred thousand years ago. Similar studies on the male-specific Y chromosome have likewise led to a "Y Adam," who dwelled in Africa at about the same time.

It should be noted, for those who would like to see a biblical analogy in these findings, that they do not refer to a specific couple, but to two unspecified individuals who are estimated, on the basis of theoretical calculations, to have been part of an initial population of about five thousand members of each sex sharing the same genetic endowment. As time went on, all but one of the same-sex lines initiated by members of this population were cut short by the failure of females to beget daughters or of males to beget sons, leaving only two uninterrupted lines, one female, the other male, going back to two individuals that most likely never have had anything to do with each other. Y Adam almost certainly never copulated with mitochondrial Eve.

According to these findings, all extant human beings are descendants from this single African branch. Contrary to the polygenism hypothesis, the bands of *Homo ergaster* and *Homo erectus* that spread over much of Eurasia one and a half million years ago all became extinct long before our times without leaving any extant descendants. Hopes that some may have survived much later were recently lit by the discovery, in 2004, on the Indonesian island of Flores, of the remains of some pygmy-like individuals, which were assumed to go back to a group of *Homo erectus* that had become stranded on the island and had undergone dwarfism, as is known to happen to isolated species under such circumstances. Doubts have since been expressed as to the reality of so-called *Homo floresiensis,* which could be no more than an abnormally developed modern human. The matter is still being debated as I write this book.

The fact that, after all these wanderings, the final step to humanity should have started from a single point in Africa, its early cradle, is remarkable. It could be mere coincidence. Or, perhaps, the core process of hominization never left the African continent, and only lateral branches have spread out over other territories.

The acquisition of language was a crucial step in hominization

Among the acquisitions that marked this final stage, a special place must be given to the capacity for speech. All mammals communicate by sounds. Some even have a rudiment of a language, with different sounds having different meanings. Vervet monkeys, for example, use different alarm calls to signal the presence of a leopard, an eagle, or a snake. However, even the most meaningful sounds emitted by mammals, including the higher apes, are no more than varieties of grunts or shrieks, nothing like the articulated sounds used for communication by human populations all over the world. The difference is not just mental.

In all nonhuman mammals, the larynx, or voice box, is situated close to the pharynx, the funnel-shaped conduit that leads from mouth to esophagus. This proximity severely limits the kinds of sounds that can be produced. Humans stand out from all other mammals by having their larynx situated distinctly lower, which endows them with their unique ability to emit the wide variety of sounds that allows articulated speech. In typical evo-devo manner, this evolutionary history is "recapitulated" in early child development. Newborns have the typical larynx-to-pharynx closeness of the other mammals and, as a result, can emit only unformed cries. In the course of the first two years of development, the baby's larynx slowly descends, and the ability to speak appears. Interestingly, in acquiring language, we lost the ability to drink and breathe at the same time. Apes and babies can do this; human adults can't.

According to cranial castings, the larynx probably started descending only in the last stage of hominization, sometime between a hundred thousand and fifty thousand years ago. This crucial event may well have signaled what is sometimes called the "great breakthrough" or the "great leap forward," the dramatic expansion of human achievements that occurred in the last fifty thousand years. In the span of a few millennia, tools of much greater sophistication, including needles and awls, harpoons, spear points, and bows and arrows, began to be manufactured, not only out of stones but also out of bones, horns, antlers, wood, and other biological products. Animal skins were prepared and converted to protective clothing. Vegetal fibers were worked into ropes and nets. Settlements became better organized on a communal basis. Shelters were constructed, and fire became routinely used for heating and cooking. Big game hunting was developed to become an efficient, cooperative activity, employing specially designed weapons. Ships were built and used, not only to follow rivers or coastlines, but even to cross wide expanses of water. Especially, artworks and jewels started to be made, and the dead began to be buried according to special rites. Awe and fear became expressed in what may have been the first religious ceremonies.

These skills were not all developed in Africa, but the aptitude to exercise them must have done so, to be spread over all parts of the world in a new wave of migrations, even more extraordinary than the preceding one. Not only did the humans of those days retrace the earlier wanderings of their ancient *ergaster* and *erectus* predecessors into Europe and Asia. They went on to the extreme Far East, even crossing over to Australia in what must have been a most hazardous passage. At the other end, they reached north Russia and walked across the land junction that spanned the Bering Strait at that time, invading the American continent long before it was discovered by Christopher Columbus. Whether they crossed the Pacific Ocean from Polynesia to South America on balsam rafts, as was done successfully on his *Kon Tiki* by the Norwegian explorer Thor Heyerdahl to substantiate this theory, is more problematic.

Cro-Magnon inaugurated modern humans

The star of this adventure is usually named Cro-Magnon, from the name of the site, in the Dordogne region of France, where thirty-thousand-year-old skeletal remains of *Homo sapiens* were first discovered in 1868. It is generally believed that the Cro-Magnons were fully developed, modern humans. Give them a shampoo and dress them in a T-shirt, jeans, and sneakers, and you could meet them on Times Square without a turn of the head. Not that they have ceased evolving. Many of the traits that distinguish the so-called human races were probably acquired by Cro-Magnons as they adapted to different parts of the world. It is believed, for example, that skin color was lost by northern Europeans, who were led by natural selection to exchange this protection against ultraviolet radiation, which was locally weak, anyway, for the more vital advantage of being able to make vitamin D, which they no longer found in sufficient amounts in their food and had to synthesize in their skin by a ultraviolet light–activated reaction. There is evidence that some traits, such as the inability of adults to digest lactose (milk sugar), which characterizes many populations, in India, for instance, were acquired even later. Human evolution is still going on today, partly influenced by the advances in medicine, which allow many unfavorable genes that would normally be eliminated by natural selection to continue being transmitted. In spite of all these local variations, humans have remained a single species, by and large sharing the same genes and able to interbreed.

What happened to the Neanderthals?

Although the Cro-Magnons represent the main heroes of our story, they are not the only ones. They were preceded by an earlier branch that sprouted before them from their common African trunk and also developed considerably, spreading to Europe and other parts of the world. Fossil remains and other traces of this line were unearthed in the early part of the nineteenth century in Belgium and other sites in Europe, but their significance was not recognized until the accidental discovery, in 1856, of some old bones by workers who were cutting a quarry out of the banks of a small river called Neander, not far from Düsseldorf, in the German Rhineland. It is intriguing that the name of this river actually means "new man" in Greek and was given to it in honor of a local musical celebrity, Joachim Neumann (German for "new man"). So much for those who believe in predestination. First mistaken for bear bones, the remains fortunately came into the hands of a local teacher, who was inspired to show them to an expert anatomist, who identified them as human, though significantly modified. All kinds of weird hypotheses were put forward to explain these modifications, until the proposal, inspired by Darwin's book, which was published just three years after the find, was made, and, eventually, accepted, that the observed features may be those of a long-extinct human ancestor. Today, the Neander Valley has become a major landmark, under its German name of *Neanderthal,* which is itself perpetuated in the name of its most illustrious denizen, *Homo neanderthalensis.*

Neanderthals probably moved out of Africa before their younger kin, settling first in many parts of Europe and the Middle East. There, they were joined later by the Cro-Magnons, with whom they coexisted for some time. Much has been written concerning the similarities and differences between the two. In particular, there has been a great movement toward rehabilitating the Neanderthals from the reputation of brutishness that has long been given to them, to the point that it has become politically incorrect to use the term in a pejorative sense.

It is generally agreed that the two groups were sufficiently close to be ranked under the same label of *Homo sapiens,* with the Neanderthals named *Homo sapiens neanderthalensis,* and the Cro-Magnons *Homo sapiens sapiens.* The artifacts and living conditions of the two groups seem comparable in many respects, except culturally. There is virtually no evidence of Neanderthal art, jewelry, or rituals, at least comparable to those of the Cro-Magnons. They did sometimes bury their dead, but with little ornament. Their brains, nevertheless, seem to have been somewhat larger than those of the Cro-Magnons. But this is not necessarily a sign of mental superiority. Brain size is only a rough index of intelligence. The development of certain specific areas, in the neocortex, for example, is a decisive factor. The shape of the skull in Neanderthals and Cro-Magnons is very different, indicating a different organization of certain centers.

A major question is whether the Neanderthals were able to speak. Casts of their lower cranium suggest that their larynx may have been too high for true speech, which could possibly explain their relatively crude cultural development. On the other hand, the shape of a hyoid bone—a horseshoe-shaped bone that surmounts the larynx—found in Israel suggests otherwise. The matter remains subject to debate.

Another unanswered question refers to the relationships between Neanderthals and Cro-Magnons. Did they live alongside largely ignoring each other, as do many kindred species today? Or, did they fight one another, perhaps even to the point that the more clever Cro-Magnons exterminated the other species? Or, on the contrary, did they fraternize and even interbreed?

We may soon have answers to some of these questions, thanks to the wonders of modern technology. Samples of mitochondrial DNA and, more recently, nuclear DNA have been retrieved from Neanderthal bones, and their sequencing has begun. Present results already allow an estimate of the time when Neanderthals and Cro-Magnons separated from the last ancestor they have in common: a minimum of half a million years, perhaps as much as eight hundred thousand years, which is at least three hundred thousand years earlier than the time when mitochondrial Eve and Y Adam (see above) started the Cro-Magnon line and when some key human traits, perhaps including speech, were acquired. These traits may thus have been lacking in the Neanderthals, unless they were acquired by convergent evolution. The DNA results also make it unlikely that much genomic interchange occurred between the two. They probably did not interbreed, or if they did, their offspring was infertile, as is the case for many hybrids, such as the mule.*

Modern humans remain the only survivors from the adventure out of which they were born

The Neanderthals disappeared about thirty thousand years ago, for reasons not yet clarified. They may have fallen victim to environmental hardships they were unable to survive. Or they may have been driven out of existence by the more successful Cro-Magnons by a mechanism variously surmised to have been extermination, forced migration to climatically unfavorable regions, deprivation of vital resources, infection with deadly diseases, or hybridization leading to sterility. Whatever happened left the Cro-Magnons as sole inheritors of this long, extraordinary saga that started somewhere in Africa some seven million years ago, when a primate line first diverged from the line leading to modern chimpanzees. After the disappearance of the Neanderthals, the Cro-Magnons went on developing over many millennia, producing remarkable cultural achievements, such as the cave paintings of Lascaux and Altamira, but continuing to rely on hunting and gathering as their main means of survival. Then, some ten thousand years ago, the advantages of planting crops, rather than gathering them, and of raising animals, rather than hunting them, started to be appreciated by some people in the Near East, perhaps also elsewhere, leading to the first permanent settlements. The rest, as they say, is history.

*As this book was going to press, it was announced that certain features found in fragments of Neanderthal DNA are shared with that of modern Asian-European humans, but not of Africans, suggesting possible late interbreeding (*Science* 328 [2010]: 680–684).

10
Making the Human Brain

Of all the wonders of life on Earth, the human brain is no doubt the most wondrous. Forming this wonder, there is a special kind of cell, the neuron. Like all other cells of the organism, neurons have a body, with a nucleus and all the characteristic structures and organelles of animal cells. But, in addition, they are specialized in the reception, processing, and emission of signals. They do this by means of two kinds of extensions, the longer axons, doing signal emission, and the shorter arborescent dendrites, doing signal reception.

The brain is constructed with neurons

In their most primitive form, neurons probably connected a sensitive spot with some motor or secretory system. As a simple example, imagine a light-sensitive spot connected to a contractile fiber: spark a flash of light, and contraction occurs. Single cells may already show reactions of this sort; but the value of neurons is that they can relay messages over distance, thanks to their extensions. Those distances may be considerable, up to a dozen feet in the nerves of large animals.

But simple relays were only a starting point. What has given neurons their astonishing power is their ability to join by connections, or synapses, that allow coordination among the connected cells. Thus, in primitive jellyfish, which are probably among the first animals to have possessed such cells, the neurons surrounding the hole that permits exchanges between the alimentary cavity and the outside world are linked into a ring, so that the muscles that command the opening and closing of the hole operate in a coordinated manner.

As animals evolved toward greater complexity, neurons increased in number and importance. It is remarkable that, of the 959 cells that compose the body of the small nematode (roundworm) *Caenorhabditis elegans,* one of the most intensely studied animals, almost one-third are neurons.

Two other neuronal phenomena took place in the course of evolution. First, neuronal cell bodies started to congregate in special centers, called ganglia, while the neuronal fibers joined into bundles, called nerves. Furthermore, as animals began to acquire a head (cephalization), ganglia tended to become grouped in this part of the body, inaugurating the beginning of a brain, which, as we know, went on growing in size and complexity throughout animal evolution.

The cerebral cortex is the mysterious site of conscience

A crucial event in this history was the development of the cerebral cortex, a thin, sheetlike structure, about eight hundredths of an inch thick, that, in humans and in higher animals, envelops the entire brain. Typically consisting of six superimposed layers of neurons interlinked vertically and horizontally by a thick jungle of intermingled connections (fig. 10.1), the cortex is the seat of consciousness. Its boundary

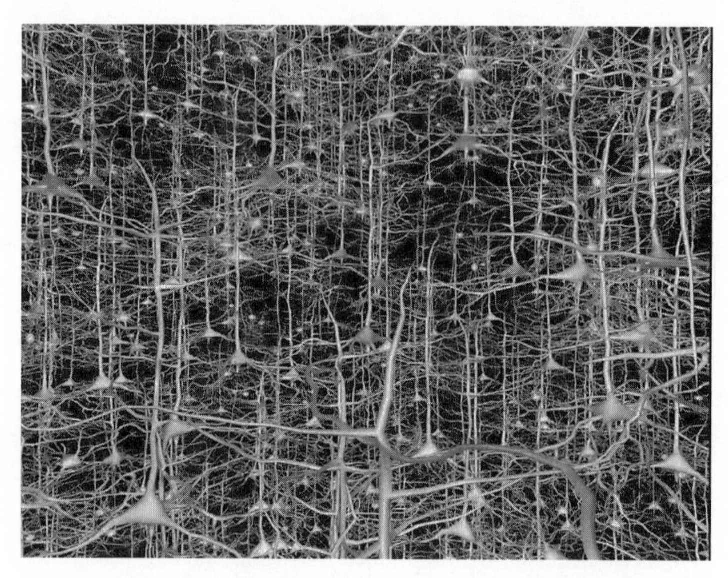

Fig. 10.1. *"Forest" of neurons in the cerebral cortex.* This remarkable computer-generated reconstruction illustrates the dense network of neuronal arborescences in a thin transverse section through the human cerebral cortex. This structure, repeated side by side millions of times, forms what is by far the most complex assemblage in the entire known universe, the generator of consciousness. Courtesy of Javier DeFelipe.

with the remainder of the brain serves as a screen between the unconscious and the conscious. On the inner side of this boundary, the complex assemblage of nerve centers and fibers that constitute the body of the brain and the rest of the central nervous system receives and processes innumerable incoming signals and sends off innumerable outgoing orders. Most of this activity takes place without our being aware of it, regulating heartbeat, blood pressure, digestion, eye movement, balance, and a host of other physiological phenomena. Some brain circuits cross the boundary and pass through the cerebral cortex. Those that do so elicit awareness, feelings, emotions, impressions, thoughts, dreams, imaginings, reasonings, decisions, the whole gamut of mental phenomena that fills our heads.

How this mind-generating machinery functions is utter mystery. Neurobiologists have accumulated a considerable amount of evidence purportedly showing that the neurons do it all—"You are just a pack of neurons," as Crick put it—with, as conclusion, that consciousness is but an epiphenomenon, some sort of aura that emanates from neuronal activity but has no control over this activity, contrary to our feeling of being in charge, which, it is claimed, is a mere illusion, a trick played on us by natural selection. The fact remains that the nature of consciousness has so far eluded objective characterization—it is a purely subjective phenomenon—and the mechanism whereby it is generated by the cortical networks is not understood. Looking at the tangle illustrated in figure 10.1 and reflecting that millions of them are joined side by side in the human cortex, with, in a single human brain, more interneuronal connections than there are microchips in all the computers of the world put together, one cannot help suspecting that this amazing assemblage is the seat of phenomena of a different order from all those described and explained by conventional physics. Perhaps it will take brains of even greater complexity to comprehend the secret of the human brain.

The manner in which this extraordinary structure arose and the nature of its earliest manifestations are also unknown. It is certainly not a human innovation. The first vertebrates already show the beginning of a cortex, and unformed conscious experiences—of pain, pleasure, fear, anger, or desire—most likely developed way down the animal line. As the animal brain increased in volume, so did the cortex in surface area. In mammals and, perhaps, in birds, conscious experiences may be quite rich. As we saw in the preceding chapter, many animals, especially those closest to humans, exhibit behaviors, such as the manufacturing of simple tools or the use of different signals in communication, that imply fairly complex mental underpinnings.

It took six hundred million years for the animal brain to reach, in chimpanzees, a volume of 21.4 cubic inches

And so, over some six hundred million years, brain size slowly increased to a volume of about 21.4 cubic inches (350 cm^3), which was the volume of the brain of the last ancestor humans have in common with chimpanzees, whose brains are about that size, whereas the cortex surface area expanded to about 197 square inches (500 cm^2), which is more than a smooth shape would allow and was achieved by the infoldings responsible for cerebral convolutions.

In the human line, it took two to three million years for the brain volume to expand from 21.4 to 82.4 cubic inches

About six to seven million years ago, that is, after animals had gone through 99 percent of their evolution, something stupendous happened, probably the most extraordinary event in the entire history of life on Earth, certainly the most momentous. In an evolutionary line that detached from the chimpanzee line and ended up leading to humans, brain volume started growing at an increasing pace, to more than three times its volume, while the surface area of the cerebral cortex expanded fourfold, producing even deeper infoldings. This dramatic history is pictured in figure 10.2.

By and large, humans are what they are and do what they

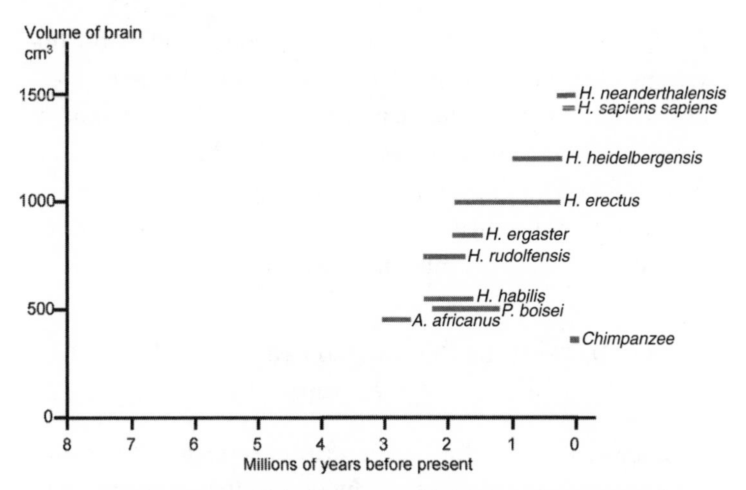

Fig. 10.2. *Brain size and the duration of existence of various prehuman groups.* The height of each horizontal line represents the brain volume of the corresponding species, as deduced from fossil cranial measurements (real measurements in the case of contemporary chimpanzees and *Homo sapiens*). The length of the lines represents the duration over which fossils of the species have so far been found. Graph constructed using data from S. B. Carroll, "Genetics and the Making of *Homo sapiens,*" *Nature* 422 (2003): 849–857.

do thanks to this epoch-making transformation. Between plucking termites with a denuded branch and splitting the atom, between calling the group together under a tree with a howl and singing *Saint Matthew's Passion* in the Sistine Chapel, the difference is one of brain size, 100 billion neurons instead of 25 billion, with a quadrupling of interneuronal connections, from 250,000 billion to 1 million billion. I neglect here the rather trivial point about absolute brain size being only a coarse measure of mental potential. Body size and internal brain structure are also important. We saw this with the Cro-Magnons, who did better than the Neanderthals, even though they had slightly smaller brains. The fact remains that absolute size *is* critical. The richness and complexity of the operations a brain can achieve depend on the number of connections among neurons, which, in turn, is limited by the number of neurons. In the present case, it is indisputable that the increase in brain volume and, particularly important, in cortical surface area, went together with greater mental abilities and hence greater accomplishments of all sorts.

The expansion of the human brain went through a number of successive plateaus

The manner in which this fateful process took place raises fascinating questions. As an introduction to the problem, let us start with some known facts, as presented graphically in figure 10.2. In this graph, each horizontal line corresponds to one of the groups mentioned in the preceding chapter. The height of the lines represents the average brain volume of the individuals in the group, as deduced from cranial measurements, and the length of the lines gives the time range over which fossils of the group have been found.

The most striking feature of this graph, in addition to the rapidity of the climb, is its stepwise pace. By all appearances, brain size "jumped" from one level to another, subsequently remaining at the new level, with little change, for very long times, exceeding one million years in some cases. In the meantime, new jumps occurred elsewhere, so that several groups with brains of different sizes often co-existed (at least in time, if not in location). Two and a half million years ago, for example, four different species coexisted—*Paranthropus boisei, Homo habilis, Homo ergaster,* and *Homo erectus*—with brain sizes ranging from 30.5 to 61 cubic inches (500–1,000 cm^3).

This picture is incomplete, depending as it does on fossils found so far. New groups may be discovered in the future and thus fill some of the gaps in the figure. But the main trend is unmistakable. It goes by way of apparently stable levels of considerable duration, separated by periods of rapid increase of which no trace has yet been found. This, incidentally, is a common finding in evolution. Links are rare or missing, probably because their existence is fleeting, in comparison with the durability of the groups that the links connect.

With evidence lacking, it remains for us to imagine the missing connections by educated guesswork. Two extreme possibilities are illustrated in figure 10.3. Both models assume, as seems likely, that the process of brain expansion was unidirectional and that regressions from an upper to a lower level did not occur.

In model A of figure 10.3, the jumps are pictured as taking place as late as the data permit, that is, at the onset of each new level. In model B, I have assumed that the steps represent horizontal branches that extend laterally from a single, uninterrupted, ascending line. To satisfy this requirement, I had to assume that some branches started earlier than the age of the oldest fossils found, which is not implausible in view of the scarcity of fossils and the role of chance in their discovery. The two models have in common rapid jumps from one level to another, followed by prolonged stabilization of the new level. The difference lies in the timing of the jumps: mostly at some stage within the lower plateau in model A; before the start of the plateaus in model B.

As far as I know, model A represents the generally held view. Its shape corresponds to that most often given in treatises for the human "tree" or "bush" (shown here lying down). Model B, first proposed in 2005 in my book *Singularities,* has not, to my knowledge, been considered before. I tend to favor it because it assumes a single genetic propensity toward bigger brains, with a number of stops on the way, whereas, in model A, a new drive toward brain expansion is initiated at each

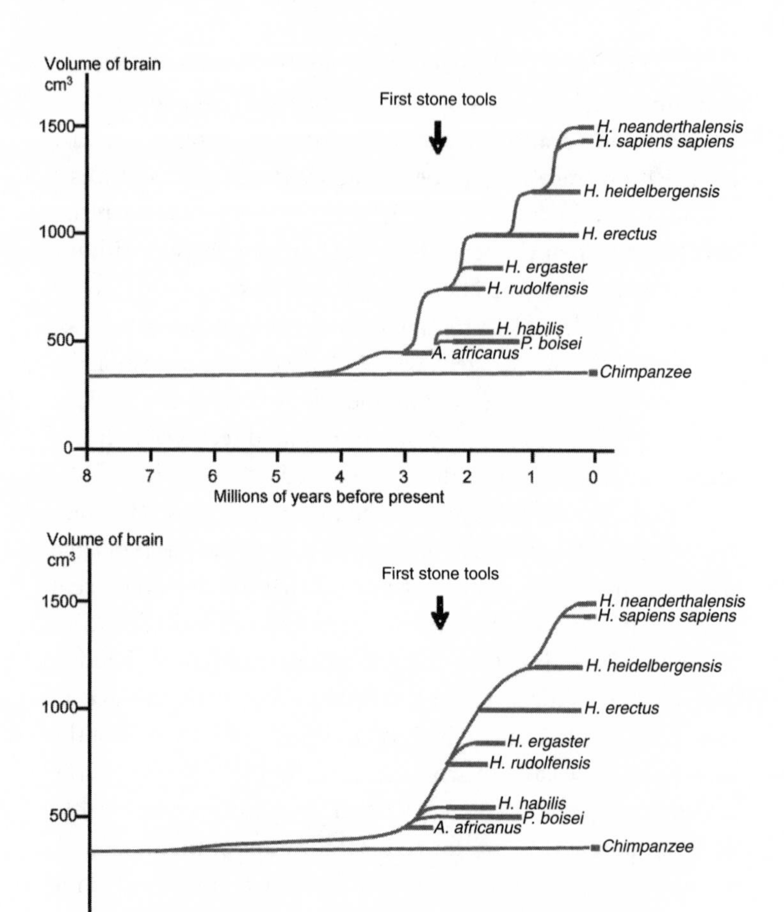

Fig. 10.3. *Two models of human brain expansion.* These two models are drawn using data from fig. 10.2. (A) Model based on the hypothesis that jumps from one level to the other occurred as late as the data permit. (B) Model based on the hypothesis that all levels extend laterally from a single ascending curve. Note that stone tools started to be manufactured by creatures with brains half the size of the modern human brain. Model B adapted from Christian de Duve, *Singularities* (New York: Cambridge University Press, 2005), 222.

level. Model B is thus the more economical of the two in terms of the number of required genetic changes. This could be a point in its favor. Indeed, an impressive aspect of hominization is the extraordinarily small number of individuals involved at any stage, as opposed to the importance of the changes that characterize the process.

We saw in chapter 9 that the group, comprising mitochondrial Eve and Y Adam, ancestral to the entire, present-day human population may have included no more than ten thousand individuals. Studies of the Neanderthal genome have suggested an even smaller number—on the order of three thousand—for the common ancestral population from which Neanderthals and Cro-Magnons diverged more than half a million years ago. That the appropriate genetic change could have taken place at each stage in such small populations is flabbergasting. It indicates that the changes must have been either exceedingly probable or very few in number. The latter condition is better fulfilled by model B of figure 10.3, but the argument is not decisive. A single tendency toward expansion that was repeatedly stifled and reawakened could account for model A.

We must leave it to the experts to decide which of the two models—or any model intermediate between the two—is more likely to be the correct one. What I wish to examine here is the manner in which the "jumps" from one level to another took place, whether in one or the other model. In my graphs, rather than connecting the levels by abrupt steps, in staircase fashion, I have rounded the angles to give S-shaped, or sigmoid, curves, which seem to me more realistic.

Mathematicians have long been familiar with this type of curve, known as the logistic curve. It reflects two opposed phenomena: one, an exponential increase such that every increment enhances the rate of the following one; and two, a

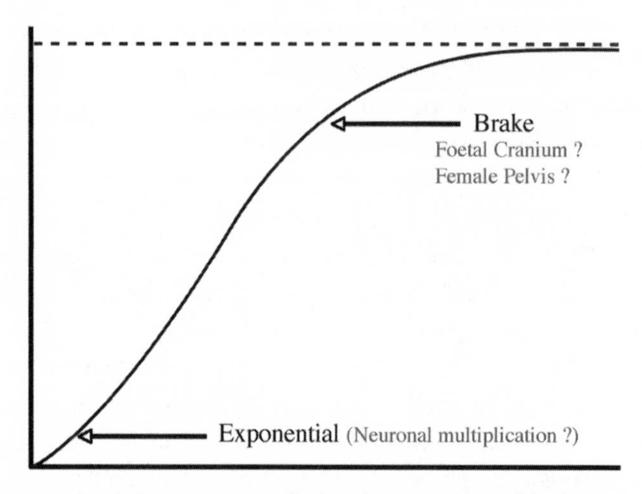

Fig. 10.4. *Logistic curve, as applied to the expansion of the human brain in the course of evolution.* The curve combines an exponential increase with a braking effect. The exponential part is attributed to neuronal multiplication and the braking effect to anatomical constraints, either the size of the fetal cranium or that of the female pelvis, or both.

braking effect that increasingly slows down the process as it progresses, until it stops at a limit value that generates an unchanging plateau. What natural processes could account for these two phenomena in the case of brain expansion? Figure 10.4 illustrates a possible answer to this question.

Exponential neuron multiplication braked by anatomical constraints probably explains the sigmoid shape of the jumps of brain volume from one plateau to another

For biologists, cell division is the iconic example of an exponential process. One cell divides into two, which divide into four, which divide into eight, sixteen, and so on. Cell division happens also to be the indispensable mechanism behind brain expansion. One is thus naturally led to the hypothesis that the exponential rise from a lower to a higher level reflects cell division. This assumption is particularly attractive, as it ascribes all the successive rises to a single phenomenon.

The simplest factor that could account for the braking effect responsible for causing the plateau is some kind of anatomical constraint. Here, several factors may come into play. First, there is cranial capacity. As the brain enlarges, it is increasingly hindered by its bony box. This leaves two possibilities. The cranium does not yield, so that only individuals in which neuronal division is genetically programmed to stop at a certain stage are allowed to survive and produce similarly programmed progeny, accounting for the plateau. Or the cranium yields to the outward pressure exerted by an expanding brain, a distinct possibility as the cranium is made of a number of plates that, in the fetus and even in the newborn, are linked by membranous connections that could indeed stretch, as required. If this happens, the drive toward a bigger brain is allowed to proceed further, until a new obstacle is encountered, generating the next plateau. A second factor that could increasingly limit brain size, as it expands, is the dimension of the female pelvis, which may oppose the birth of young with bigger heads. A possible way of overcoming this kind of obstacle would be by way of a change in the brain's developmental program, so that the brain achieves a lesser degree of maturity in the womb and completes a greater part of its maturation after birth.

Expansion of the human brain was limited by the size of the female pelvis and by the degree of immaturity at birth compatible with survival

There are clear indications that all three factors have played a role. Cranial capacity has obviously increased. Except for humans and chimpanzees, the brain volumes in figures 10.2 and 10.3 are all derived from cranial measurements. On the other hand, the female pelvis, although it may have adapted in some measure to this change, has apparently offered considerable resistance to the birth of babies with bigger brains, with, as a consequence increased immaturity at birth, or *neoteny.* This is a typical human feature.

Delivery in humans is excruciatingly painful, probably at the limit of what is bearable, whereas it seems to cause relatively little discomfort in other mammals. Development of the human brain inside the womb obviously takes place until the size of the head has reached the utmost limit compatible with passage through the birth canal. Even so, the brain is still at a very immature stage and continues to undergo outside the womb a developmental process that, in other mammals, is largely completed in utero. A human baby is entirely helpless at birth and will need many months to reach a stage where it can match the autonomy of, for example, a newborn foal. In all appearance, natural selection has, in humans, favored birth at an increasingly immature stage as the price to pay for a better brain. This brings us to the history of our genes, the subject of the next chapter.

11
Shaping Our Genes

We are, like the rest of the living world, largely products of natural selection. Our genes are there because, at some stage in evolution, they happened to be useful to the survival and reproduction of their owners or, at least, were not sufficiently harmful for their owners to be eliminated. Some 98.5 percent of those genes existed in the last ancestor we have in common with chimpanzees and were gained at some stage in the long pathway that led from the first forms of life present on Earth more than three and a half billion years ago to the last bifurcation that separated the hominid primate branch from the chimpanzee branch, some seven million years ago. Those genes account for all the properties we share with chimpanzees. We owe what makes us specifically human to the remaining 1.5 percent. It sounds little, but it is still a lot, a genetic text of about forty-five million "letters," a pretty fat book.

Hominization involved an astonishingly small number of individuals

In trying to identify those genes and to explain their selection, we immediately encounter a challenging difficulty. The number of individuals involved at any stage of hominization must have been very small, probably never exceeding a few thousand, if that many. A wide range of mutations is not expected to occur at any time in such a limited population, and the probability that an appropriate mutation will be included in the lot is correspondingly reduced. Thus, assuming an average lifetime of twenty years and a total population of five thousand, the number of new births and, therefore, the number of genetic variants offered to natural selection would be on the order of 250 per year, which appears very small. This number is, however, no longer insignificant if duration is taken into account. Over fifty millennia, for instance, which is less than one-hundredth the total time taken by the hominization process, the number of variants exceeds ten million.

Also important, the population was probably divided into small bands of some thirty to fifty members, closely inbred and related by kin. In such a situation, whatever favorable mutation was offered had a good chance of being efficiently exploited, as it would quickly spread within the group and provide its members with an edge over other groups that did not enjoy the same advantage. We shall see the key importance of this feature.

The fact remains that the chances of evolutionary success were bound to be greater if favorable mutations had a high probability of taking place and if the number of required mutations was small. It is interesting, in this respect, that some of the main differences that have been found between the human and chimpanzee genomes affect genes that influence other genes by way of transcription regulation (see chapter 6). Thus, changes in only a few such master genes could have had wide-ranging effects. Also suggestive, in relation to brain expansion, is the observation from human pathology that brain size may be controlled by very few genes. Two kinds of inborn microcephaly (small brain) have been traced to the mutation of a single gene.

Let us now look at the long pathway of hominization within the framework of natural selection. This pathway may be divided in very approximate fashion into three successive stages, involving different genetic changes and selective factors (see figs. 10.2 and 10.3).

Hominization probably started with bipedalism, which was selectively advantageous in the local terrain

First, until about three million year ago, the main element was the conversion from quadrupedalism to bipedalism, accompanied by only a moderate increase in brain size. Then, covering the next million years, comes a period of rapid brain expansion, associated with the making of stone tools of increasing sophistication. Coming finally are the great migrations out of Africa, which took place in two waves: one, starting some two million years ago and invariably ending in extinctions; the other, launched some half a million years ago and bringing forth, first, the Neanderthals, who also disappeared, and, later, the conquerors of the planet and only survivors of the adventure, *Homo sapiens sapiens*.

Considering first the conversion from quadrupedalism

to bipedalism, one is struck by how many anatomical changes must have accompanied this conversion. Legs became longer, arms shorter; feet became adapted to walking, hands to grasping; the spine, pelvis, and shoulders all underwent changes that facilitated an erect posture and the corresponding gait; the position of the head with respect to the rest of the body was modified; muscles, nerves, blood vessels, and viscera followed the skeleton in evolving to allow new kinds of movements and adapting to different weight distributions. The number of genes that had to be modified for all these adjustments to be achieved cannot be guessed but must have been substantial.

Most of the changes involved, however, happened after commitment to bipedalism had already occurred; they represent improvements, added slowly over several million years, that benefited creatures that had converted to bipedalism. This conversion itself was the decisive factor. Many of our primate cousins walk on two legs intermittently or even regularly, so the final tip-over could have depended on very few genetic changes. What selective advantages may have resulted from the adoption of bipedalism is a question that has prompted much speculation and discussion. Displacement from the forest to the savannah, as assumed by the "East Side Story" (see chapter 9), may have played a significant role. Freeing the hands may have been another major advantage.

Brain expansion dominated the second major stage of hominization

What about the next stage of hominization, highlighted by brain expansion and tool-making? Here, an environmental explanation of natural selection seems much less probable than for the first stage. One would expect a more effective brain and more skillful hands to be selective assets under any environmental condition. Furthermore, there seems to be no reason to suspect that this stage of human evolution involved major environmental upheavals. For all we know, it could have taken place mostly under more or less unchanging conditions, in the semiarid regions of northeast Africa. Remember the traces of steps preserved in Laetoli ash during more than three million years (chapter 9).

The dominant factor under such stable environmental conditions could have been competition among bands for the richest hunting and gathering grounds, especially if foods were scarce and difficult to come by. Competition among males for the most desirable females could have been another factor. Here we encounter what may have been a crucial stage in our history, with natural selection reinforcing in our genes two traits that were there before but became increasingly important as the ability to act purposefully improved: solidarity within groups and hostility among groups, especially manifested by males. We shall see in the following chapter to what extent this legacy still affects human behavior today.

The vagaries of environmental change probably guided the migrations that characterized the third stage of hominization

What caused the third stage of hominization and first drove *Homo erectus* and his contemporaries out of Africa, and later impelled *Homo sapiens* to inaugurate a new series of migrations, is not known. Search for food, perhaps in the wake of migrating herds, probably was a factor. Once the migration started, environmental challenges no doubt again played an important role in natural selection. As the migrant groups encountered different climates and milieus, different traits were retained or rejected by natural selection. We have seen, for example, how loss of pigmentation may have been favored in northern areas, where the weak UV radiation emitted by a pale sun became an asset, rather than a liability. Environmental differences may likewise account for other, less well understood changes that differentiate the major human ethnic groups. Remarkably, in spite of these changes and of long periods of isolation from each other, the different groups—the so-called human races of earlier treatises—never evolved to the point of no longer being able to interbreed. We form a single species.

Hominization: Chance or necessity? Summit or stage?

Looking back over the fabulous story of our origins, we can't help asking a number of questions. Was the birth of humankind the outcome of an extraordinary combination of circumstances, of a unique conjunction between an improbable genetic accident and environmental conditions that happened by chance to be conducive to exploitation of the accident? Or, on the contrary, had the evolution of the primate brain reached a point such that only a flip was needed to trigger a process that was, so to speak, written in the genes at the time and ready to be initiated? In the latter event, what could have been the triggering flip? If it had not happened, could some other event have started things going? Last question, has the process reached a summit? Or could it continue and one day lead to "superhumans" with mental capacities distinctly superior to ours?

We obviously have no answers to these questions. Concerning the first one, all that can be said is that the pace of hominization, as it can be quantitatively estimated from that of brain expansion (see fig. 10.3), suggests an irresistible process that, once set in motion, proceeded at a growing, almost autocatalytic rate, repeatedly overcoming the obstacles that opposed its progress. This looks more like a case of evolutionary maturity, waiting only for a triggering flip, than like an extraordinary stroke of luck, raising the question of the nature of the triggering flip. Could it have been a critical mutation of a gene controlling the size of the brain? Or an environmental accident, such as the replacement of forest by savannah assumed in the "East Side Story"? Or both? There is no way of knowing, but one is tempted to suppose that, if events had been different, the necessary triggering flip would nevertheless have occurred, so impressive is the apparently obligatory character of the process, once initiated.

As to the last question, the present situation allows no prediction. One can simply say that there seems to be no objective reason for assuming that hominization has reached an unsurpassable summit. This question will be further examined below (see chapter 14).

12
The Cost of Success

They numbered a mere three thousand about half a million years ago, in the heart of Africa, when the Neanderthals left them to go their own way. There were little more than ten thousand of them, some three hundred thousand years later, when mitochondrial Eve, Y Adam, and their congeners set off on their final stretch towards *Homo sapiens sapiens;* an estimated five to ten million, thinly scattered over a good part of the world, when the first durable human settlements were created ten thousand years ago. Since then, they climbed to about half a billion in 1600, one billion around 1800, two billion in the 1930s, four billion in the 1970s, and more than six and a half billion today. If nothing changes, they will exceed nine billion in 2050. The exponential increase of the human population predicted by Malthus is dramatically manifest, even exceeded (fig. 12.1).

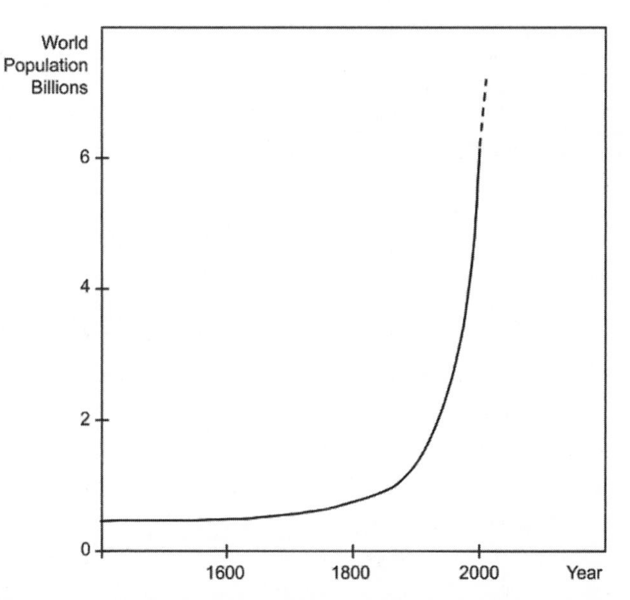

Fig. 12.1. *Human population expansion in the last centuries.* The rise has grown to hyperexponential in the last two centuries, but may be starting to slow down in recent years, initiating a logistic curve.

Taking advantage of the powers of their brains, humans have proliferated beyond all measure and exploited a major part of the planet's resources for their own benefit

In their expansion, humans have invaded every part of our planet, from below sea level to the highest mountaintops, from tropical forests to frozen steppes and ice fields, from lush savannahs and prairies to the driest deserts, from the Earth's surface to the depths of oceans, the air above us, and even the Moon and distant space. Unlike other living species, they have not achieved their successes by developing appropriate physical adaptations; they have done it with their intelligence.

They have used the powers of their brains to subjugate the world, domesticating much of the biosphere to obtain their food and fuel, as well as the raw materials they have used to manufacture their clothes, shelters, ships, weapons, hunting gear, fishing implements, and countless other devices. They have turned mud into bricks, clay into pots, sand into glass, rocks into cement, mortar, and concrete. They have quarried stones for building houses, temples, and roads. They have extracted minerals from the Earth to produce copper, iron, tin, aluminum, and other metals, which they have used as such or converted into a powerful set of new materials, including bronze and, especially, steel, which has transformed the world. With these products, they have built cities, harbors, highways, bridges, factories, and machines. They have invented engines to work for them, and exploited natural energy sources, including the fossilized remains of ancient biospheres, to fuel those machines. They have discovered electricity as a transportable form of energy, generated at one end from heat, falling water, sunlight, wind or, even, the atom, and converted at the other end into heat, lighting, cooling, mechanical work, and other useful actions. They have woven a dense transportation and communication network around the planet. They have, by the wonders of chemistry, transformed simple materials into all kinds of artificial substances, including plastics and drugs. They have made telephones, radios, television sets, computers, robots, and all the other electronic devices that support our technological age. With the help of some of these tools, they have explored the depths of the universe, the nature of matter, the secrets of life, and the mysteries of the mind. They have succeeded in mastering nature to the point of engineering life almost at will and of unleashing the huge amounts of energy stored in atoms. They have created music, art, literature, philosophy, and religion, which are, with science, the pillars of human civilization.

There is no story like it. After more than three and a half billion years of slowly evolving life, almost imperceptibly giving rise to microbes, plants, fungi, and animals of increasing complexity, a single species among the many millions born from this development has suddenly exploded into overwhelming world dominance within no more than a few millennia, most of it within the last few centuries, roughly the equivalent of the last minute if life had started one year ago.

The history of humanity is a perpetual succession of wars and conflicts

Amazingly, these collective achievements have been accomplished despite perpetual strife or, sometimes, because of it. Human history, from the time records have become available and, probably, before that, has been an endless succession of battles, wars, crusades, conquests, invasions, massacres, colonizations, and enslavements of various kinds. Most of the monuments that adorn our marketplaces and village squares are representations of triumphant emperors, kings, generals, admirals, liberators, and other warriors. Or, usually more modestly displayed, they are tributes to the millions who have lost their lives for the sake of victory—or defeat—as celebrated by the poet Horace, who, two thousand years ago, encouraged the young Romans with the obscenely famous verse: "Dulce et decorum est pro patria mori" (It is sweet and decorous to die for one's country).

Even in peace, competition has remained, directly or by proxy, the most widely, passionately, and, sometimes, violently practiced form of entertainment, with arenas, stadiums, and sports fields replacing bloody battlefields as outlets for our thirst for combat. Business and commerce also are strongly competitive, with few holds barred. So is politics. Even academe has its battles for tenure, prizes, and peer recognition; art and music have their contests. All these manifestations, whether bellicose or peaceful, are rooted in the depths of our being. The warring instinct is embedded in human nature.

The inordinate evolutionary success of the human species has been acquired at the expense of a severe deterioration of living conditions on Earth

In a different register, human history has been a savage, collective exploitation and wanton destruction of natural resources, living and nonliving, for the sake of immediate benefit with little or no regard for long-term consequences. The living world has become impoverished, species are being lost every day, energy and other resources are nearing exhaustion, the environment is deteriorating, pollution is everywhere, climate is changing, natural balances are threatened. Especially, human beings are being crushed by their own number. Overcrowded cities are spawning increasingly lawless suburbs. Waste is accumulating in and around them, straining the capacity to deal with it. Vast areas are witness to the struggles of destitute populations trying to survive under unlivable conditions. In spite of the advances of medicine, deathly epidemics are more menacing than ever before. Conflicts, exacerbated by economic disparities, nationalisms, and fundamentalisms, are raging in various parts of the world. The specter of a nuclear holocaust has become thinkable.

If it continues in the same direction, humankind is headed for frightful ordeals, if not its own extinction

For all these reasons, the exponential pace of human expansion may be about to flatten into a logistic curve, with the limit being set by the finite dimension and resources of planet Earth. This enforced flattening, if it occurs naturally, is bound to be achieved at the cost of enormous human suffering through famine, deprivation, disease, environmental assaults, and internal strife.

13
Original Sin

Viewing the somber image just depicted with the eyes of a biologist, I find a single culprit: *natural selection.* I use the word "culprit" metaphorically, of course—no guilt is involved—but the term is not entirely inappropriate. Natural selection, this all-powerful driving force of biological evolution, has privileged in our genes traits that were *immediately* favorable to the survival and proliferation of our ancestors, under the conditions that prevailed there and then, with no regard for later consequences. This is intrinsic to the process of natural selection, which sees only the immediate present and does not foresee the future. Note that I refer here only to genetically acquired traits. I leave out traits that were acquired by cultural evolution and transmitted by education. These will be considered later.

Natural selection has indiscriminately privileged all the personal qualities that contribute to the immediate success of individuals

Human traits retained by natural selection have proved extraordinarily fertile. Without attempting an exhaustive inventory of those traits, which is beyond our scope at present, I find among them a number of individual properties, including intelligence, inventiveness, skillfulness, resourcefulness, and ability to communicate, which we owe to the remarkably performing brains we have acquired in the last few million years, qualities that have generated the fantastic scientific and technological achievements responsible for our success.

But the selected traits have also included selfishness, greed, cunning, aggressiveness, and any other property that ensured immediate personal gain, regardless of later cost to oneself or to others. The worldwide financial crisis that hit like a storm in the fall of 2008 illustrates in particularly dramatic fashion how such traits still persist in today's world. On the other hand, natural selection has little favored qualities, such as long-term prevision, prudence, a sense of responsibility, and wisdom, which would have proved advantageous only in the long run. Their fruits would have appeared too late for that.

Natural selection has privileged traits favoring cohesion within groups and hostility among different groups

On the collective level, natural selection has privileged traits, such as solidarity, cooperativeness, tolerance, compassion, and altruism, up to personal sacrifice for the common good, which are the foundations of human societies. But selection of those traits has generally been restricted to the members of groups. The negative counterpart of those "good" traits has been defensiveness, distrust, competitiveness, and hostility toward the members of other groups, the seeds of the conflicts and wars that landmark the entire history of humanity up to our day.

It probably all goes back to the time when small bands of prehumans competed for the best resources offered to them by the African forests and savannahs, perhaps even much earlier, as group solidarity is a characteristic of many animal societies. At first, the group was defined by kinship—with the accent on family, clan, or tribe—as it had to be for the traits to become genetically imprinted. Later, the group expanded to encompass shared territories, shared needs, shared interests, shared privileges, shared beliefs, shared values, shared prejudices, shared hatreds, shared anything that could serve to unite "*us against them.*" Domineering nationalisms and religious fundamentalisms play this uniting role today.

Natural selection has not privileged the foresight and wisdom needed for sacrificing immediate benefits for the sake of the future

The genetically determined search for immediate profit, whether individual or collective, also explains our irresponsible exploitation of natural resources and lack of concern for the nefarious consequences of our activities, the effects of which are now threatening the future of our species and of much of the living world. Anything that goes beyond the immediate future, whether relating to our retirement, our life expectancy, the fate of our children and grandchildren, or the date of the next elections, to mention only a few familiar deadlines, hardly preoccupies most of us.

All those facts are known and abundantly denounced by the media. What I, as a biologist, have wished to emphasize in this book is that they are the outcome of traits that are *inborn*, written and sustained in our genes by natural selection. They were useful in the past, at a certain stage of our evolution, but have become destructive. They are a natural burden that we assume at birth. This defect of human nature has not escaped the sagacity of our ancestors.

Original sin is none other than the fault written into human genes by natural selection

The wise men of the past knew nothing of DNA or of natural selection. But they knew enough about heredity to write the history of humankind in terms of successive generations going back to the first parents. They also knew enough about human nature to perceive in it a fundamental flaw inherited from the parents and transmitted from generation to generation. They thus imagined, in order to explain this hereditary stain in terms of notions that were familiar to them, the marvelous myth of original sin, situated in the nostalgic site of paradise lost. Not to yield to utter despair, they have invented the idea of salvation, the redeeming act that would save humanity from its downfall. This myth still inspires today the beliefs, hopes, and behaviors of a good part of humankind. That is why calling natural selection the "culprit," as I have done in the beginning of this chapter, was not entirely inappropriate, except that no culpability, in the proper sense of the word, is involved. There is no Eve to blame, no serpent, no dangerous fruit, only natural selection, necessarily blind, mindless, devoid of foresight and responsibility.

The only possibility of redemption from the genetic original sin lies in the unique human ability to act against natural selection

Less romantic than the account of Genesis, the proposed notion has the merit of being founded on reality. Instead of calling on a hypothetical redeemer totally beyond our control, it confers on humanity itself the power and responsibility to erase the original stain or, at least, to counteract its effects. We are indeed, of all living beings on Earth, the only ones that are not slavishly subject to natural selection. Thanks to our superior brains, we have the ability to look into the future and to reason, decide, and act in the light of our predictions and expectations, even against our immediate interest, if need be, and for the benefit of a later good. We enjoy the unique faculty of being able to act *against natural selection.*

The problem is that, in order to do this, we must actively oppose some of our key genetic traits, surmount our own nature. The last part of this book will be devoted to this challenge.

IV
The Challenges of the Future

Introduction

The wisdom of the ages enjoins us to draw lessons from the past in preparing the future. This is what I have attempted to do in this book, going back to the most distant past to draw the lessons that could help humanity respond to the deathly menaces that weigh on its future. Such is the object of this last part. Here, I consider seven options, which are not necessarily mutually exclusive.

14
Option 1: Do Nothing

Our first option is to do nothing and let nature take its course. In that case, there can be little doubt about the outcome.

If nothing is done, humanity is headed for disaster

The signs are unmistakable. Humans have unwillingly and, even, unwittingly, by unbridled pursuit of immediate benefits, endangered their own survival to such an extent that, for all we know, the point of no return may already have passed. Under the mercilessly indifferent law of natural selection, this blind, suicidal course can only continue until it is brought to a halt by its own consequences. Natural selection has no foresight.

The extinction of humankind, if it occurs, will be due, not to its failure, but to its success

If humankind were to become extinct, we would only be suffering the same fate as all hominid species that have preceded us, but with a major difference. What caused the demise of our hominid predecessors is not known, but it most likely involved their inability to withstand some external menace, such as a geological catastrophe, a climatic assault, food scarcity, infectious disease, or competition with a more successful species. Whatever the cause of the extinction of our predecessors, it most likely was associated with some kind of *failure* in the face of a natural hardship. In our case, extinction would be due to a uniquely different reason: *inordinate success.* The problem is in our genes. We are the products of natural selection, victims of a genetic quirk that has given us enough intelligence and skillfulness to conquer the world but not enough wisdom to husband the fruits of our victories.

Could a "superhuman" species succeed the human species?

If our species were to disappear, the question arises whether history might one day again repeat itself. Could a new sigmoid curve one day be initiated from the plateau we occupy (see fig. 10.3) and lead to a new, higher plateau, characterized by a brain even bigger and more effective than ours? As we saw in chapter 11, there seems to be no objective reason for assuming that hominization has reached an unsurpassable summit. On the contrary, the apparently irresistible character of this process, once set on course, would rather prompt one to believe the opposite and to assume that it can but progress further if given the opportunity. This condition might, however, not be easy to achieve.

A first requirement would be the appropriate genetic mutations. Our past history, which was accomplished with a remarkably small number of individuals, suggests that the genetic propensity for brain expansion, which has not ceased during more than six million years, in spite of several obstacles, would not be hard to acquire and could even still be latent in our genome. What may be more chancy are the accompanying anatomical changes necessary to make further brain expansion possible. Considering the manner in which the cranium grows, from bony plates joined by membranous connections, little change would be needed to allow the cranial box to house a bigger brain. Birth would be a greater problem. If we were right in assuming that the size of the human brain is limited, on one hand, by the degree of immaturity at birth compatible with survival and, on the other, by the size of the female birth canal, either one or the other or both of these parameters would have to change. The necessary modifications could affect the offspring, allowing birth at a stage of lower brain maturity, followed by a longer period of extrauterine maturation. Or they could affect the female pelvis and allow the passage of newborns with bigger heads.

Given the necessary mutations, prevailing conditions would have to be such that the mutations would spread among a small inbred group, which, benefiting from the advantages provided by a superior brain, would then inaugurate a further rise in brain size, until some obstacle once again interferes to curb the exponential curve into a sigmoid curve that ends at a new plateau. It is difficult to see how such conditions of reproductive isolation could obtain in today's world. This does not mean, however, that they will not be realized some time in the future. As an extreme possibility, hardly to be wished but conceivable, the dire conditions created in the last throes of humankind could lead to a situation where a small cluster of advantaged survivors emerge, thrive, and propagate their advantage.

Whether in this or in some other way, it is possible, even probable, that something of the kind will happen some day in the future and that beings with bigger brains than ours will one day exist. There is no reason, except hubris, to assume that we are the crowning achievement of evolution and that the human brain has reached the ultimate in terms of possible development. Considering our mental limitations and our past history, our present existence may very likely be only a transient, intermediate phase in an adventure that is far from finished. Plenty of time is left for such continuation.

Life has up to five billion years left before Earth becomes incapable of harboring it

According to astronomers, the Earth still has some five billion years left—more than its entire past—until the Sun, its energy stores exhausted, explodes into a red giant, annihilating the essential conditions for life on our planet or, even, elsewhere in the solar system. Some experts believe that the Earth may, for one reason or another, become physically unable to bear life long before that; but even the most pessimistic among them give life on Earth a minimum of one and a half billion years to go. Within the framework of the history of life, these are immense stretches of time, amply sufficient for a new brain-improving or, even, brain-creating adventure to be launched, possibly evolving toward greater success than our species has achieved.

Whatever happens, life will not disappear until it is forced to. Its ability to adapt is such that only physical conditions totally incompatible with its persistence could succeed in eradicating it. However poor the biosphere we leave behind, it will recover and, perhaps, reach new heights, before the Earth becomes totally unable to harbor life.

What could happen in a brain even more developed than the human brain?

Taking as a reasonable hypothesis that individuals with bigger brains will one day exist, it is intriguing to speculate on the nature of the mental experiences these individuals would enjoy. Alas! By definition, this is no more possible for us to do, with our 82.4-cubic-inch brain, than it would have been for Lucy, with her 24.4-cubic-inch brain, to apprehend the thoughts of Einstein, Beethoven, Michelangelo, or Shakespeare, or even those of us ordinary humans of our day. I like to imagine that the owners of bigger brains and, especially, of wider and denser "forests" of cortical neurons (see fig. 10.1) will understand more clearly, feel more deeply, and approach more intimately the "ultimate reality" that I tend to perceive behind the appearances accessible to our limited human faculties. Perhaps they will be closer to understanding the mystery of consciousness as a bridge toward this reality.

As premise for this chapter, I have adopted the assumption that our future is to be shaped exclusively by natural selection, as was the case for all our predecessors. Should this be so, it may be safely predicted that whatever future is awaiting us will be reached only at the cost of extreme hardships, a simple extrapolation from the problems we are facing today. But this is not an absolute necessity.

With the advent of humankind, evolution has reached a point where it is no longer a slave to natural selection

While it is true that humanity is most likely not the crowning achievement of evolution and may one day be followed by creatures with greater mental powers, its advent still represents a watershed in the history of life. For the first time, after more than three and a half billion years, evolution has passed a threshold in its dependence on natural selection, with the production of beings endowed with the intellectual means to understand nature, including their own, the practical ability to turn this knowledge into acts capable of changing the course of events, and the awareness of what we call moral responsibility for the effects those acts may have on the human circumstance. Unlike the rest of the living world, we are not necessarily the toys of blind, irresponsible natural selection. We are, to some extent, masters of our fate. The following chapters will examine some of the options open to us.

15
Option 2: Improve Our Genes

If there is a flaw in our genetic makeup, the most straightforward way to address the problem is to correct the flaw. This project is not new. Measures aiming at improving the hereditary endowment of humanity were proposed in the nineteenth century under the name of "eugenics," literally the pursuit of good genes.

Eugenics has become a dirty word

This term was conceived, with the project it implies, by a cousin of Darwin, Sir Francis Galton (1822–1911), who initiated intelligence testing and defended the view that psychological characteristics are hereditary and that society should defend itself against undesirable traits by selective breeding, if not by stronger measures against the genetically unfit. First favorably received in certain conservative and racist quarters of Victorian England and the United States, eugenism prompted a number of discriminatory measures, condemned today, with as summit of ignominy the horrors of the Nazi regime. Since then, anything smacking of genetic discrimination has become socially and politically abhorrent.

The most shocking aspect of the original eugenic project is the distinction it made between bearers of "bad genes" and bearers of "good genes," and, even more, the means it proposed to eliminate the former or, at least, prevent them from procreating and so propagating their flaws. But another form of eugenics that does not suffer this objection can be envisaged today; it aims at eliminating or correcting certain unfavorable genes *we all have in common,* genes that were once adaptive, even essential to our survival, but have become increasingly maladaptive. If these genes are now dysfunctional although retained by natural selection, perhaps we should use our brains to correct them before they do us in. We may well have the means to do so, at least potentially.

Cloning opens the way to directed evolution

It started in 1996 in Scotland, at the Roslin Institute, near Edinburgh, where a small team headed by Ian Wilmut first succeeded in cloning a sheep, the now world-famous Dolly. Since then, the procedure has been successfully repeated on more than a dozen different mammalian species, including mice, rats, cows, pigs, horses, cats, and dogs. Success with humans has been claimed but not substantiated. What is the procedure?

First, what does "cloning" mean? Derived from the Greek *klon,* twig, the noun "clone" is defined as a group of organisms formed asexually from a single ancestor. Bacteria typically develop as clones. So can some plants, starting from a single cutting. Until Dolly, however, mammals had never been cloned. Actually, calling Dolly a clone is not strictly correct, as Dolly is an individual, not a group. But the word is now sanctioned by

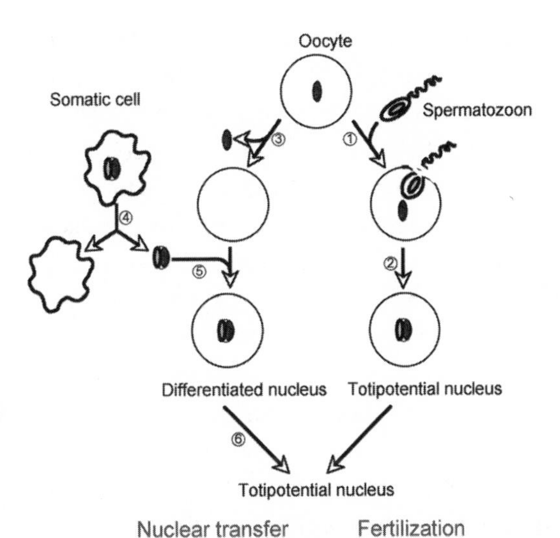

Fig. 15.1. *Cloning by nuclear transfer, as compared to fertilization.* On the right, the normal mechanism of fertilization: (1) two gametes, oocyte and spermatozoon, each with a single genome (haploid), join to form (2) a totipotential fertilized egg cell possessing two genomes (diploid) and a cytoplasm supplied almost exclusively by the oocyte (see fig. 5.5). On the left, nuclear transfer: (3) an unfertilized oocyte is enucleated; (4) a somatic (body) cell with two genomes (diploid) is enucleated; (5) the diploid nucleus of the somatic cell is inserted into the enucleated oocyte, to produce a cell with a diploid, differentiated nucleus and the oocyte's cytoplasm; (6) the differentiated nucleus is "deprogrammed" under the influence of the oocyte's cytoplasm and is returned to a totipotential state. Either cell is now ready for implantation.

usage and refers to a nuclear-genomic "carbon" copy of an individual obtained without sexual reproduction.

On paper, the method whereby this result is achieved is simple, but it is very delicate in practice (fig. 15.1). One takes an unfertilized oocyte, removes its nucleus, and replaces it with the nucleus of a differentiated cell, from skin, liver, or mammary gland, for example. The renucleated oocyte is then implanted into a female body, where, under proper stimulation and if all goes well, it develops into a normal individual with a nuclear genome identical to that of the donor of the inserted nucleus, as in an identical twin, except that, unlike the identical twin, the individual would have a different mitochondrial genome (which comes from the oocyte). This quasi-identical twin of the owner of the transferred nucleus is what we now call a clone.

Note the change in chromosome number caused by this procedure. We saw in chapter 5 that oocytes, like spermatozoa, have a single set of chromosomes; they are haploid. When an oocyte and a spermatozoon join in fertilization, they give rise to a diploid cell, with two sets of chromosomes, one set from the oocyte, one set from the spermatozoon. All the other cells of the body derived from the fertilized egg are likewise diploid, with the exception of germ cells, which undergo reduction to a haploid state in the course of their maturation. Cloning, therefore, means replacing the haploid nucleus of an unfertilized oocyte with the diploid nucleus of some body cell; it is a means of providing an oocyte with a diploid nucleus without fertilization (see fig. 15.1). For the rest, it involves the same manipulations as are widely applied in the procedure known as in vitro fertilization (IVF).

A key difference between nuclear transfer and fertilization lies in the potential of the nucleus. In fertilization, the resulting diploid nucleus is totipotential; it can, by specific activation of certain genes and silencing of others, give rise to any of the differentiated nuclei characteristic of the different cell types found in the body. This is what happens in the course of the successive cell divisions whereby the fertilized egg develops into a complete organism. In cloning, this totipotential nucleus is replaced by the already fully committed nucleus of some differentiated body cell. One might expect to see such a renucleated egg cell multiplying into a clone of the cell that provided the nucleus. Surprisingly, this is not what happens. Under the influence of the oocyte cytoplasm, the developmental clock turns back to zero in the transplanted nucleus, which undergoes "deprogramming" to the totipotential state. As a result, the inserted nucleus recovers its lost youth.

This astonishing return to virginity by a transplanted nucleus was first observed in the 1970s, in amphibians, by the British biologist John Gurdon, who is the true "father" of animal cloning. Shortly afterward, a Swiss investigator claimed to have accomplished the same result on mice but later had to retreat in some confusion—perhaps unjustly in view of what is now known of the hazards of the technique—because his findings could not be reproduced. As a consequence of this unfortunate incident, mammalian cloning fell into disrepute, until Dolly was produced by Wilmut and his team, who obviously belonged to those who "didn't know it couldn't be done and so went ahead and did it."

What can cloning be used for?

Cloning can be used for several purposes (fig. 15.2). One form, called *reproductive* cloning, aims at creating a younger, quasi-identical twin of the donor of the transplanted nucleus. This method can be used, for example, to perpetuate valuable stock.

In so-called *therapeutic* cloning, the resulting early embryo is sacrificed, to provide totipotential stem cells that are kept for possible future use in the repair of damaged tissues in the donor of the nucleus. The advantage of this technique is

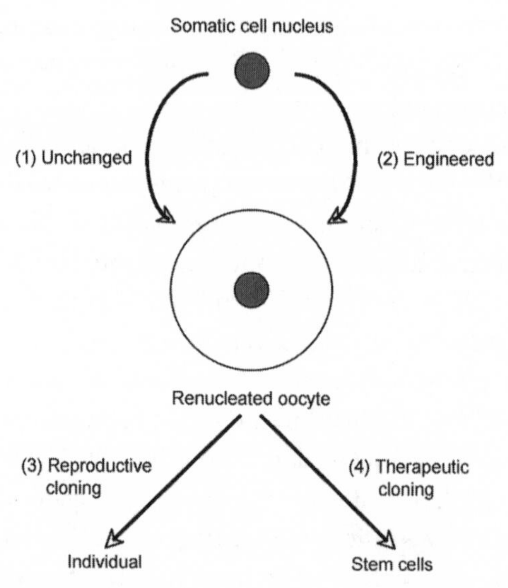

Fig. 15.2. *Different kinds of cloning.* The somatic cell nucleus destined to be transplanted into the enucleated oocyte is inserted as such (1), or after modification by genetic engineering (2). The renucleated oocyte is either implanted to develop into an individual (3, reproductive cloning) or used for the generation of stem cells (4, therapeutic cloning). Combination of steps 2 and 3 is used to create genetically modified (transgenic) animals.

that, because of their genetic kinship with the cells of the donor, such stem cells are not immunologically rejected by the host, as would be the case with foreign grafts.

In a particularly important form of cloning, which could be called *engineering,* the transplanted nuclei have been subjected to some genetic modification, leading to the creation of genetically modified organisms (GMOs), also called "transgenic." This procedure is now widely used in research, where it has allowed a host of important discoveries. There are also many industrial applications of this technique, used, for instance, to generate animals that manufacture valuable human proteins in their milk or to endow animals with new, useful properties, such as the ability to subsist on new kinds of foods, enhanced productivity, or lower harmfulness to the environment. Thus, pigs have been equipped with a bacterial enzyme that renders the animals able to digest an important phosphorus-containing component of their food that is normally excreted unbroken and is a major source of pollution of streams and lakes in the neighborhood of pig farms, where it favors eutrophication, an excessive proliferation of algae that stifles other forms of life. These animals have been called "enviropigs" because of their beneficial effect on the environment. Genetically modified plants are also produced on a large scale, by a different procedure, which I describe in chapter 18.

Human cloning provokes heated ethical debates

So far, no authenticated instance of human reproductive cloning has been reported. But this is due mainly to ethical constraints that most countries impose on such attempts. There is no reason to suspect that the technology would not be applicable to human cells. Certainly, this possibility has already inspired innumerable conjectures and debates. Creating a younger copy of oneself or replacing a lost child have been cited as possible applications of human reproductive cloning, a procedure that is, at present, prohibited by most legislatures around the world.

Human therapeutic cloning is allowed in many countries but fiercely opposed in others, including the United States, where prolife advocates condemn the destruction of a potential human embryo inherent in the technique. Because of this opposition, an immense research effort is aimed at obtaining stem cells without cloning. Possible sources are umbilical cord blood or, if some genetic commitment is acceptable, differentiated tissues, most of which are now known to contain pluripotential (though not totipotential) stem cells. Special hope is put in a recently achieved procedure claimed to turn back the clock in a fully differentiated cell by the modification of only four genes.

As to engineering cloning, its envisaged human applications are restricted to the medical correction of some specific gene defect already identified. The use of this technique to produce "designer babies" has, however, been much evoked, hotly discussed, and mostly rejected.

Production of designer babies would require considerable improvement of present cloning methods. In today's state of the art, the percentage of failures remains too high to allow an ethically acceptable use of human material. The number of "failed" embryos and their fate would create unmanageable situations. Should safe procedures someday become available and should the ethical ban against human reproductive cloning be lifted, this technique could conceivably be used to engineer genetically modified humans, by knocking out or otherwise mutating certain genes or by inserting new ones.

The main problem, if such manipulations were to become possible and ethically acceptable, would be the choice of the genes to be manipulated or inserted. We would need to know much more than we know now about the genetic control of human qualities. This is a major unsettled issue. Experts even disagree on the extent to which certain characteristics are genetically determined, let alone agreeing on more specific relationships. It does, however, seem probable that complex psychological traits or talents do not depend on single genes. It is unlikely that one could engineer the production of a new young Mozart or Einstein or Martina Navratilova by simple genetic manipulation.

Even if the necessary knowledge of genetics should one day be available, one still would have to agree on what qualities to aim for. If some central authority were in charge, the threat of a *Brave New World* would loom. In order to avoid this dangerous difficulty, some have proposed formation of a genetic "supermarket," in which prospective parents would choose the qualities of their offspring à la carte. This idea may seem grotesque, but it could arguably be viewed as an improvement over the present lottery, in which the genetic makeup of children is decided by pure chance among a huge number of combinations of parental genes. Remember that, because of the genetic reshuffling that takes place in the course of germ cell maturation, oocytes and, especially, spermatozoa come in a large number of different genetic varieties. The genome of a fertilized egg depends on whichever of the millions of spermatozoa contained in an ejaculate succeeds in entering the available oocyte first. Substituting reasoned choice for such a blind game could be seen as desirable.

Whatever happens, humanity will not be saved by cloning

Whether such manipulations will ever be attempted or accomplished cannot be predicted at the present time. They certainly are not among the measures that can be contemplated today as a solution to the present pressing problems of humankind. Even if one knew what to do and how to do it, the question would remain: Who should benefit from the improvement? Changing six billion individuals hardly seems feasible. Starting with a small group destined to become the new *Herrenvolk* is too reminiscent of Nazism to even be thinkable today. Letting selection be based on the ability to pay for the very expensive procedures involved, as proposed by the American biologist Lee Silver in his book *Remaking Eden* (1997), would be consistent with laissez-faire, though not with most people's concept of democracy or social fairness. It will be up to future generations to make the appropriate decisions if they are ever able to produce viable genetically modified humans. The problem does not presently exist, but it may well some day.

16
Option 3: Rewire the Brain

Correcting the flaw is not the only solution to our genetic predicament. Genes can be overruled by education. This message comes to us from neurobiology, which tells us that some among our most decisive traits are *epigenetic.* I use this adjective in the original meaning given to it in the 1930s by the British evolutionist Conrad Waddington (1905–1975) to qualify traits that are not genetically transmitted and are acquired later in life, under genetic control but in response to outside factors. In recent literature, as we have seen in chapter 8, the noun "epigenetics" designates a new kind of genetics involving transmissible traits that are not encoded in DNA sequences but accompany the DNA. The two definitions are thus contradictory in terms of heredity.

The wiring of the brain is an epigenetic phenomenon

The wiring of the human brain is a striking case of epigenetics in the Waddington sense. Only the general features of the brain are genetically determined. Its detailed wiring is superimposed upon the genetic blueprint, it is epigenetic. It couldn't be otherwise. The human brain contains about one hundred billion neurons, each of which is connected with some ten thousand other neurons, adding up to about one million billion interneuronal connections. Our genome contains only about three billion bases, not nearly enough to determine that many interneuronal connections, even if—which is hardly conceivable—each base should code for a connection. This fact opens hopes for the future, offering a way for us to escape the possible fatality of genetic determinism. This escape is particularly meaningful, as it concerns the brain, the organ through which we make decisions and perform actions.

The way the wiring of the brain is established epigenetically has been elucidated by the investigations of Jean-Pierre Changeux, in France, and Gerald Edelman, in the United States. According to these scientists, growing neurons continually send out projections in all directions. Acting like "feelers," these projections, upon chance encounters with each other, form transient connections that quickly come apart again if they are not used. If a stimulus repeatedly goes through such a connection, it becomes stabilized into a synapse. Both of these scientists have stressed the analogy between this mechanism and Darwinian selection. Chance offers a vast array of possible connections among which a small number get selected by use. Edelman calls it "neural Darwinism."

The implications of this epigenetic process are critical. The human brain is shaped to a large extent by the impulses to which it is exposed during the first years after birth, even, perhaps, while in the womb. The process continues all life long, by education, training, and learning. Even an old brain can make new connections. But the first years are crucial. Children deprived of contact with other humans for the first years of life are permanently stunted psychologically.

Education starts in the cradle

These findings have a profound significance for the topic of this chapter. If we wish to take advantage of the plasticity of the brain to counter the defects imprinted in us by natural selection and escape the tyranny of our genes, we must start in the cradle and continue afterward in the early educational environment to which a child is exposed, in a nursery, kindergarten, or primary school, or at play with peers or parents and other grown-ups. In other words, for children to be changed, their parents and teachers must first be changed.

This looks like an impossibility. It requires parents and teachers to be changed in adulthood. Furthermore, in order to be effective on a global scale, the change would have to affect hundreds of millions of largely illiterate parents, as well as their children's elementary schoolteachers. Instant change, over a single generation, is clearly impossible. Evolution, step by step, toward the desired situation is a more realistic ambition. Even if started on a small scale, such a movement, if sufficiently contagious, could snowball into worldwide enlightenment.

Many such attempts have been made in the course of history. Think of pacifist religious groups, the Amish and Quakers, the advocates of nonviolence, from Christ to Buddha to Gandhi, conscientious objectors, the communes and flower children of recent years, "make love, not war," and so on. None of these movements has snowballed. Some have even degenerated into violent defense of nonviolence. But this is no reason for despair. With increasing awareness of the disaster we are facing if we do not change course, future initiatives could meet with greater success.

Political and, especially, religious leaders are particularly well placed to propagate the recommendations the world needs

Even a large-scale movement is not unthinkable. History shows that mass indoctrination of adults by single individuals is possible. Political leaders have done so repeatedly, but mostly within the confines of national borders and with aims that were far from pacifistic. But some philosophers and, especially, religious leaders have managed to influence huge masses across national boundaries. They, more than anybody else, are in a position to help spread the epigenetic changes needed to save the world.

17
Option 4: Call on Religions

Churches have been the foremost dispensers of education to the young during much of history and they still play a major role in this crucial process in many parts of the world. Even in the United States, where public schools are run by local lay authorities, Churches still exert important influence, by way of school boards and other supervising bodies. Their involvement is weaker in Europe, where religious neutrality of the public school system is strictly enforced and respected; but religious influence remains significant, through private and, sometimes, even state-supported schools.

Churches are also much involved in adult education, through sermons, homilies, and other exhortations they deliver to their congregations at their gatherings. They are thus ideally placed to spread the epigenetic corrections to our genetic heritage that are urgently needed to save the world from irreparable, human-inflicted damage.

Churches could play an exceptional role in saving humankind

In a way, this is what religions traditionally have striven for. Invention of the myth of original sin, with its attendant need for salvation, may well represent the earliest human appreciation of our fundamental, inborn behavioral defects and the necessity that they be corrected. In particular, the Christian message of love, peace, tolerance, and forgiveness is exactly what is called for in our troubled world. Historically, this message lies at the root of much of civilization, and it is shared, beyond ideological differences, by many of the adepts of the major religions, as well as by numerous nonbelievers. Unfortunately, the message is in danger of being lost, stifled by the tendencies it was meant to correct. Churches have not escaped the genetic "original sin" that plagues the whole of humanity. Several factors inherent to their nature prevent them from playing the role one would wish them to carry out.

Religions are founded on beliefs, not on rational thought

A major weakness—and, paradoxically, also strength—common to most religions is their reliance on *belief*, that is, unquestioning faith in affirmations by an authority that has no other legitimacy than its own claim to hold the truth, supported by a powerful ability to convince. Belief in a creator God, for example; in the veracity of the biblical record; in resurrection of the dead, with, in the end, eternal bliss (or doom). Belief in animism or reincarnation; in the efficacy of prayer or in the power of certain sacramental gestures and rituals. Belief, especially, in the authority, sometimes, even, infallibility, of a hierarchy that has invested itself with the right to decide what is true. Acceptance without proof of such claims flies in the face of reason and can no longer be upheld in our modern world, trained, since the days of the Enlightenment, to exercise what the French philosopher René Descartes (1596–1635) has called "methodic doubt." Yet, a major fraction of human beings living on Earth today go on adhering to a system of beliefs of one kind or another and to reject rationality, or even to oppose its teaching, on religious grounds.

This paradox has an explanation. As pointed out by many contemporary thinkers, the need to believe, especially within the context of a group, is ingrained in human nature, probably carved into it by natural selection because those populations that believed in something had a greater chance of surviving and producing progeny under prevailing conditions, whatever the plausibility, or lack of it, of the object of the belief. Historically, religions have catered to this need in a uniquely effective fashion, by proposing powerful myths that directly appeal to the sense of wonder and awe humans have experienced in the face of the mysteries of the world ever since the days when their brains first became capable of generating such sentiments.

Religions have fulfilled this function for almost all of human history. Only in the last two millennia or so have, first, philosophy and, later, the sciences, started to pursue this quest by a new approach based on the use of reason, logic, observation, and experimentation, under the guidance of intellectual rigor and honesty. This new form of search has gained few adherents so far. By and large, a majority of believers, often egged on by their leaders, have failed to follow the scientific approach, in spite of the spectacular practical achievements—nuclear power, space travel, genetic engineering, the cure and prevention of many diseases, to mention only a few—it has made possible. In much of the world, the power of Faith remains supreme and religions maintain a largely unquestioned authority. A perverse aspect of this situation is the feeling of certainty that goes with it.

Many religions present themselves as defenders of the truth

Another major obstacle preventing religions from carrying out a truly educational function is the claim by many of them to be, almost by definition, the holders of the supreme Truth. They may make token gestures toward tolerance, ecumenism, universal brotherhood, and the like. But the very fact of believing implies rejection of different beliefs. It leaves no room for compromise or rational discourse. Conceding that your neighbor could be right implies admitting that you could be wrong.

As an aggravating circumstance, such a feeling of certainty almost inevitably generates missionary zeal, the wish to convert those who think otherwise. Such attempts are mostly done by peaceful means in the world today. In the past, the "propagation of Faith" often degenerated into repressions, tortures, conflicts, and wars, the bane of our genetic heritage and the very curse religion could be expected to free us from. The Catholic Church, although it has become pacific in recent times, has behind it centuries of cruel inquisitions and merciless persecutions, crusades, and other bloody expeditions that aimed at evangelization but used the bringing of salvation to heathens as a pretext for killing, plundering, and conquering. Other Christian Churches have had their share of religious wars. Strifes between Protestants and Catholics have shaped much of European history and have torn Ireland up to very recent days. Religious intolerance, added to economic hardship, has been behind several of the migrations that gave birth to the United States.

For the two other major monotheistic religions, it is not just their past that is weighing them down. They are engaged at this very moment in deadly, religiously inspired wars. Islamic factions fight each other mercilessly in Iraq, Afghanistan, Lebanon, Palestine, and elsewhere. Jihadists direct their attacks against their archenemy, Israel, and against much of the Western world, exploiting religious beliefs to preach hatred and fanaticism, up to persuading young men—and now women—to sacrifice their lives for the privilege of killing "infidels."

The Jews, in spite of—perhaps partly because of—having suffered so much from oppression, persecution, and extermination, with the horrors of the Holocaust as culminating ordeal, have, with the creation of a religious state deriving its legitimacy from a millennia-old biblical past, planted the seeds of a new tragic conflict. I realize that this is a complex situation, which is variously appreciated by Jews all over the world and even in Israel. It cannot be compared to the kind of worldwide terrorism waged by extremist Islamic factions. But the core Zionist objective of "recovering Judea and Samaria" and reclaiming its biblical roots cannot be denied.

To blame religion alone for these many excesses would be grossly oversimplified. Economic pressures, political influences, and nationalistic passions all play a role. But the "Gott mit Uns" (God is on our side) rallying cry pervades many conflicts, right down to the "war on terror" of recent years.

It must be added that this criticism concerns mostly religions issued from the Bible. On the whole, Eastern religions are less inclined to proselytism and are more pacific. I know them too little to say more about them.

Religious doctrines have a major impact on ethical directives

A particularly delicate domain is that of ethics, long the preserve of religions and still based largely on religious or philosophical principles. Many present-day controversies, on topics such as homosexuality, abortion, euthanasia, stem-cell research, and other sensitive issues, illustrate the impact of doctrine on ethical judgments. The ban against as innocuous a practice as contraception, still unyieldingly enforced by some major religions, shows how far this influence may be pushed.

Science has little to say in this domain, except for what the French biologist Jacques Monod (1910–1976) has called the "ethics of knowledge," respect for truth, with utter honesty, rigor, and integrity; a powerful prescript, but restricted to the intellectual sphere. On many other issues, science does not have the capacity to pass judgment. It is unable to decide what is good and what is bad.

Science may have little to say on ethical problems, but it is not voiceless. It can clarify numerous discussions with its findings—stem cells are a case in point. It can also render service by more clearly defining the likely outcomes of certain policies. Problems related to the environment, for example, call for ethical decisions of fundamental importance. It is urgent that religions state their position on this issue, which concerns the very future of humanity and that of the living world of which it is a part and on which it is dependent. The ethics of environmentalism are yet to be defined. What are our duties with respect to planet Earth? A number of thoughtful religious leaders are now addressing this question and the results are promising. There will be more about this in the next chapter.

Hopes for a future life could hamper efforts in favor of present life

Preoccupation of several major religions with a hypothetical afterlife complicates the question of our planetary duties. They teach, in more or less explicit fashion, illustrated by the examples of saints and other blessed individuals, that gaining a worthy place in the other world is more important than leading a happy life in this one, to the point that deprivation, suffering, hardship, even martyrdom and, in some cases, murder by suicide are sometimes made into virtues and celebrated as worthy prices to pay for a bliss to come, promised to be all the more beatific, the greater the sacrifices accepted to attain it. Prospects, such as the deterioration of our planet, the destruction of the biosphere, or the extinction of humankind, which loom on the horizon as major menaces weighing on the future, have long been viewed by many religions—and still are by some today—as inevitable outcomes, not to be opposed but to make ready for. Doomsday, the Apocalypse, Armageddon, the Last Judgment, Avenging Angels, and other end-of-the-world scenarios are prominent parts of religious mythologies.

Are religions to be fought or can they be enlisted?

What are we to conclude? Objectively, there are few reasons to be optimistic. Speaking only of the Church I know best, the Catholic Church, it is certainly less bellicose and less intolerant than it was in the past. But it remains dominated by a clerical hierarchy often more concerned with fine points of dogma or with secondary details of sexual behavior than with the real problems of humanity, more attentive to faith than to facts. And yet, it could play an exceptional role. The pope, with his unique prestige and authority over one billion faithful, is one of the few people in the world who could single-handedly alter the future course of human affairs. Numerous pontiffs have taken advantage of this privilege to attempt to influence the world, but without escaping from the dogmatism that surrounds and supports their authority. The other Bible-based religions could likewise greatly help to improve the global human circumstance but all too often remain hamstrung by their own doctrinal biases.

Is that a valid reason to abandon all religion? To this question, the most vocal and militant defenders of secularism and atheism respond by an emphatic YES. One may, however, wonder whether this response is opportune. In my opinion, it has two defects. First, it is inefficient. It seems mostly to reach those who are already convinced and hardly touches the others or, even, proves counterproductive, prompting them to rise up in arms in defense of their cherished beliefs without even giving the opposition a fair hearing. In addition, rejecting religions ignores the numerous benefits they have accomplished and go on accomplishing throughout the world.

Churches are engaged in many valuable activities

Churches do much more than propagating creeds and issuing ethical directives. They have done, and continue doing, a lot of good in the world. They play a major role in education, health care, social work, help to the disabled, assistance to the disadvantaged, and other charitable undertakings, especially in the Third World. They certainly deserve encouragement and support in such worthy endeavors. In addition, they also serve valuable social functions. Chapels, churches, and cathedrals may have, like temples, mosques, and synagogues, lost much of their primary raison d'être as "places of worship," but they continue to offer, as they have done for times immemorial, conducive settings for reflection, meditation, and contemplation, as well as for such collective activities as teachings, moral exhortations, happy celebrations, sorrowful leavetakings, and other group occasions of deep human significance. Such settings deserve to be preserved even if the manifestations that take place in them are slated to evolve as a function of changes in societies. As I can testify from personal experience, a funeral service in a beautiful church, with the sound of movingly sung Fauré's *In Paradisum*, conserves its special poignancy even though there is no paradise for the dear departed to enter.

What should we do?

How to preserve the many beneficial social activities religions accomplish within their own sphere without giving up the critical rationalism on which the scientific endeavor rests? Such, for me, is the crucial question facing us today. My answer, for what it's worth, is: by each sacrificing all that can be sacrificed "without losing one's soul," if I may be permitted this religious expression.

Religions no doubt will have to make the biggest sacrifice, by yielding to the uncontested discoveries of science in all the fields of knowledge that depend on it. This implies, for religions, a profound and painful reappraisal of many of their teachings and submission to the scientific manner of addressing problems, instead of demanding the uncritical acceptance of affirmations that are either unsupported or supported only by writings made several millennia ago by men who may have been wise but knew little about the world and our place in it. Such reappraisal, if pursued deeply enough, could even provide a new and more satisfactory response to the religious hankering imprinted in human genes, by emphasizing, as so many great scientists have done, from Darwin to Einstein, the feeling of empathy and reverence with respect to what, in my book *Life Evolving* (2002), I have called "*ultimate reality,*" inspired by our new understanding of the world.

On their part, scientists should prove as open-minded and accommodating as demonstrated facts allow. Thus, for example, while denying biological evolution on religious grounds is clearly inadmissible, claiming on scientific grounds that God does not exist is equally objectionable, for such an affirmation—as well as its opposite—is scientifically undemonstrable. Both are a matter of personal belief, fiercely argued and rationalized on both sides, but, in the last resort subjective, not objective. It is likewise possible, if not commendable, to respect in others a number of beliefs one doesn't share, albeit on scientific grounds, but that cannot be proven false. What is not tolerable, on the other hand, is the claim by certain religions that they hold the truth without providing proof of this affirmation or, even, knowingly and deliberately negating scientific acquisitions. Respect for faith must stop at what is demonstrably inadmissible.

Ethics without doctrine is possible

Even though most religions are directly based on systems of beliefs, their ethical precepts can be adhered to without subscribing to the substance of the beliefs. Many confirmed atheists lead as moral lives as the most devout of churchgoers; even more moral in some cases, judging by the recently uncovered rash of pedophilia among the Catholic clergy.

It is possible to follow the recommendations of Jesus

without believing him to be the son of God, miraculously conceived without sexual intercourse and resurrected after dying on the cross, to "ascend" in full bodily shape to some mysterious, heavenly abode, which many keep locating in the sky in spite of the discoveries of cosmologists, where the good among us are due to join him for eternity after our own death.

It is likewise possible to obey the teachings of Buddha without believing his spirit to be reincarnated at each generation in the body of a young chosen child and without seeing our own life as the manifestation of some immaterial principle that can move up and down the animal scale depending on the quality of the effort we make to free ourselves from our material constraints. It is possible to be a disciple of Moses or Mohammed without crediting them with the privilege of God-given authority.

In other words, while it is true that religions often base their moral directives on creeds, such a foundation is not indispensable. Notions such as equality, liberty and fraternity, social justice and human rights, rationality and critical thought, defended by secular humanism can suffice.

The dialogue between science and religion is desirable but difficult

The last few years have witnessed an increasing number of initiatives aimed at establishing a dialogue between representatives of the scientific and the religious worlds. So far, such initiatives have enjoyed only moderate success, because this dialogue cannot, as would be the case in politics, economics, or other fields of societal interest, end in a compromise. This is not possible in the present case. When there is contradiction between what science *knows* and what religion *believes*, there can be no compromise; religion must yield. But, for this to happen, the distinction must be made between authentic scientific knowledge and merely an opinion held by a majority of scientists, or even all, for reasons that are science-inspired but not conclusively demonstrated. As we have just seen, biological evolution belongs to the former category; the nonexistence of God to the latter.

Another problem is that such attempts at a rapprochement often seek unavowed objectives, such as inducing the admission by some scientists that "science does not explain everything," with, as outcome, a legitimization of certain vacuous "theories," such as intelligent design.

In a different vein, in an impressive setting in the heart of the Vatican, the Pontifical Academy of Sciences gathers some eighty members chosen among all disciplines and all parts of the world, without distinction of nationality, gender, and even, to some extent, philosophical opinion. It includes a certain number of overtly Catholic scientists and a few clerics—among them, until his accession to the throne of Saint Peter, the present pope, elected at the time when he was Cardinal Ratzinger—but also Protestants, Jews, Muslims, occasionally adherents of other religions, together with an appreciable number of agnostics and unbelievers, assembled under the joint umbrella of scientific excellence. In addition to plenary sessions, which are interesting through the contacts they allow among representatives of all major scientific disciplines coming from all over the world, which national academies do not allow, this unique body organizes a number of specialized meetings, bringing together experts, mostly nonmembers, to debate in complete freedom some of the most sensitive issues of our times. The only shortcoming is that the influence of the Pontifical Academy of Sciences on the Church magisterium may not be as significant as one would wish (although more may happen behind the scene than one suspects).

Possibly more important than such scholarly exchanges is the questioning by an increasing number of believers of the teachings of their own Church. This may be more of a European phenomenon than an American one. Certainly, in the circles that are familiar to me, in France and Belgium, for example, there are several signs that this is beginning to happen. In the Catholic Church, many sincere believers admit that the so-called articles of faith must be taken with a "grain of salt" and that many ethical rules imposed by the Church are to be interpreted as wishful recommendations rather than rigid laws. A growing number of so-called progressive theologians publicly agree with such views and engage in open-minded discussions with freethinkers. The opposition between the two camps is clearly dwindling. A similar trend is observable in the Church of England. Evangelical denominations in the United States seem less disposed toward such free interchanges.

Slow reappraisal from the inside may possibly prove more effective in bridging the gap between science and religion than aggression from the outside, more likely to evoke resistance rather than submission. The militant atheism proselytized by authors such as Richard Dawkins, has, in spite of the success of their books, caused hardly a dent in the armor of even the most open-minded believers, who cannot help being shocked by the virulence with which their most sacred beliefs are attacked. No, if religions are to be changed, it will happen only from within, most likely by some sort of grassroots movement starting from the base, rather than by an unlikely conversion at the top. There is little doubt that peaceful interaction will prove more effective in the long run than open warfare.

Religions, through their influence, and the sciences, through their knowledge, must urgently collaborate for the salvation of humanity

However desirable the reforms advocated above, the differences between science and religion are not about to be resolved. They are too fundamental for that. But the two can join forces, with science supplying the hard factual information on issues such as natural resources, biodiversity, climate change, energy supply, pollution, health and disease, demography, nuclear energy, cloning, genetic engineering, and other technological innovations; and with Churches providing their facilities, clergy, membership, and worldwide influence to dispense education and promote appropriate actions. They could, if only their leadership took the initiative, launch a new crusade on behalf of a modern kind of salvation, destined to rescue humanity from the consequences of its genetic "original sin."

18
Option 5: Protect the Environment

Religions are not the only organizations capable of influencing human behavior on a worldwide range. Environmentalism has become a major power in this respect. This is a very recent phenomenon.

Protecting the environment is a very recent human concern

From the time they first appeared up to the near-present, humans have exploited and polluted the world without scruple, in pursuit of immediate benefits. Not that they are to be blamed. They were not aware of the harm they were causing. The explorers and conquerors who laid the basis of the colonial empires and first settled the American continent treated the world as a source of unlimited bounty, with no regard for the consequences of their plundering.

In the nineteenth century, the advances of science and technology and the triumphs of the Industrial Revolution born from these advances were acclaimed with unbounded enthusiasm, almost devoid of concern for their harmful effects. The world belonged to us; its resources were there to serve us. If a problem should arise, science will have the answer. Even in the first half of the twentieth century, as I well remember, there was little worry about the long-term outcome of our actions.

This attitude was partly linked to the political climate of the time. The fact that progress had certain drawbacks did not go unnoticed. The smoke and soot generated by heavy industry were notorious. But those nuisances were accepted as a valid price to pay for the gathered benefits, especially as they affected mostly the worker populations grouped around the factories, whose welfare did not much concern the ruling classes. Only little more than fifty years ago were voices first raised to denounce the excesses of human exploitation and defend the environment on a planetary scale. This is a crucial "first" in the history of humankind.

An interest in nature did awaken earlier, but of a different nature. Inspired by the romanticism of the time and by the back-to-nature message of the Swiss-French philosopher and writer Jean-Jacques Rousseau (1712–1778), this new movement was directed at preserving in pristine, "virgin" state, together with their flora and fauna, some picturesque, mostly out-of-the-way natural sites. In the United States, a similar trend, prompted by the vivid reproductions of plants and animals by John James Audubon (1785–1851) and by the works of such writers as Henry David Thoreau (1817–1862), George Perkins Marsh (1801–1882), and John Muir (1838–1914), the founder of the Sierra Club, led to the creation of the first national parks, inaugurated in 1872 with Yellowstone National Park. Also dating back to those times is the American Forestry Association, founded in 1875 to combat the devastation of

American forests by wholesale logging following the conquest of the West, an innovative organization, perhaps the first to be explicitly aimed at restoring natural resources destroyed by human activity.

In those days, however, the main attention was given to distant, mostly exotic places. The local countryside and, especially, the urban environment attracted little solicitude. The situation started truly changing after the Second World War. Eloquent warnings, though sometimes disputable in certain details, by such visionaries as Aldo Leopold, Rachel Carson, Margaret Mead, Barry Commoner, René Dubos, James Lovelock, Paul and Anne Ehrlich, Peter Raven, Daniel Janzen, and E. O. Wilson in the United States, followed, in Europe, by the Club of Rome and others, began to be taken seriously, leading to formation of powerful groups of environmental advocates, some of which organized into political parties, especially in Europe. Today, the Greens have gained a significant place in world affairs.

Ecology has penetrated daily human life

The major threats facing us are now acknowledged, and measures against them are discussed and sometimes even implemented. The greenhouse effect and resulting climate changes are increasingly recognized. Measures to curb carbon dioxide emissions have been devised and, to some extent, embodied in law. Much has already been accomplished to save endangered species and to protect the environment against chemical pollutants. Efforts are made to exploit renewable energy sources, with solar panels and wind-powered generators springing up in many parts of the countryside and along seashores. Projects to harness natural imbalances, such as tides, marine currents, and temperature inequalities, are contemplated. Much investment has also been made in development of environmentally friendly fuels and energy-saving engines, lightbulbs, refrigerators, and other appliances. More attention is being paid to the expenditure of energy in food production and some improvements have been made. Waste is increasingly recycled. In a general fashion, energetic costs and pollution effects have become significant parameters in any technological initiative. Thanks to the growing use of contraceptives, population expansion is slowing down, though not as much as it should.

All these accomplishments are recent and deserve to be commended and encouraged, the direct result of a newly acquired awareness of our planetary responsibilities. Unfortunately, they are not without drawbacks.

Ecology has become the source of major controversies

The main problem is that the recommended measures have a price. They demand from the citizens, especially in industrialized countries, that they modify their habits and sacrifice all sorts of comforts that are taken for granted by those who benefit from them and are seen as enviable by the others. In addition, some of those measures threaten powerful, entrenched interests and menace considerable sources of potential profits, thereby generating the inevitable admixture of politics into the debate. To cite only one example, the interminable disputes on global warming and on curbing carbon dioxide emissions illustrate the complexity of the conflicts generated by this kind of situation. The puny results of the conference on climate, held in Copenhagen in December 2009 under worldwide coverage and with great expectations, highlight these difficulties.

Another difficulty is that scientists do not always agree

on the nature of the problems or on the interventions those problems call for. Such disagreements are almost unavoidable in view of the partly conjectural aspect of the discussions. The future projections on which recommendations are based are uncertain and debatable. Scientists, being prudent and critical, as required by their profession, feel obliged to underline those uncertainties, a fact the opponents of certain measures and the experts on which they call do not fail to exploit.

The resistance of certain political and economic groups against envisaged measures are not the sole causes of discord. The environmentalists themselves—a minority of them, at least—also bear some responsibility. Regrettably, the movement they have launched has too frequently let itself be dominated by a vocal minority that conflates justified concerns with ideological and political positions that are foreign to science and, sometimes, even opposed to it. Environmental protection, this eminently respectable, essential, and just aim, is increasingly linked with systematic hostility against technological innovations, whatever they may be, and even against the scientific enterprise in general, accused of being mainly responsible for the ills that are being fought, with no regard for the powerful aid it can provide to this fight for environmental improvement.

There is a surprising difference between the United States and Europe in this respect. In the United States, new technologies have mostly been accepted with little opposition. In Europe, environmentalism has become intimately mixed with political objectives. Certain extremist groups even go so far as to amalgamate technological realizations with the capitalistic system, which they hold responsible for the harmful consequences of globalization and do not hesitate to combat by any method, including violence and lawlessness. Even in its more pacifist manifestations, ecological activism too frequently bases its stands on assertions of poor scientific credibility and on arguments that are more passionate and irrational than objective and rigorous.

Nuclear energy: pro or con?

The debate about nuclear power illustrates this problem. Nuclear power has drawbacks. Its installations are subject to accidents that, even though rare, can be major disasters, as shown by the Chernobyl catastrophe. There is also the serious problem of nuclear waste, whose safe disposal has not yet been satisfactorily solved. Nuclear power stations need a lot of cooling water and contribute significantly to the warming of waterways and resulting damage to aquatic life. The vulnerability of nuclear power stations against terrorist attacks is also a threat. There is, in addition, the risk, particularly acute in some parts of the world, of a drift from civilian to military applications of nuclear technology.

Despite these many drawbacks, the reality of figures must be faced. Human energy needs are enormous and growing every year. So far, most consumption occurs in industrialized nations. But there is every reason to anticipate the day when each household in China, India, and other developing parts of the world will have—or demand—a refrigerator, a television set, a dishwasher, and other appliances, if not one or two automobiles. The planet's stores of fossil fuels are finite and will be exhausted in a few centuries. In the months that have gone by while I finish writing this book, the world has increasingly been shaken by the vertiginous rise in the price of oil and its equally sudden collapse because of severe self-inflicted damage to the global financial system. Whatever these fluctuations, oil supply is bound to dry up in a not-too-distant future. Coal and natural gas will soon follow. In addition, fossil fuels have serious defects. Their retrieval is not environmentally harmless or risk-free, and their use produces carbon dioxide, contributing to global warming. Whether renewable, nonpolluting energy sources will suffice to cover growing needs is far from certain. Few, if any, entirely reliable projections have been made. It is remarkable, in this connection—and deeply deplored by many Greens—that one of the first advocates of ecology, James Lovelock, the father of the "Gaia" model, has recently admitted, albeit reluctantly, that the energy requirements of the world will not be met without nuclear power.

All these factors should be included and weighed carefully in a rational, public debate. But such care and rigor are hard to come by. The fear of radiation, its invisibility and treacherous effects (inducing cancer, for example), its immense destructive capability when accidentally unleashed, have all been exploited to inflame the general public against nuclear power. In several European countries, including my own (Belgium, a pioneer in the development of nuclear power, on which it depends for more than 50 percent of its electricity), the decision has been made to abandon nuclear power, even though no adequate alternatives have been proposed to meet future needs.

The tide seems to be turning in this respect. In the United States, where nuclear energy has long suffered from neglect, nuclear plant construction is about to resume on a large scale. Even in Europe, voices are increasingly raised against the ban imposed by several countries on the development of nuclear power. In Belgium, the decision has been taken to prolong three nuclear power plants beyond the date foreseen for their closing.

The systematic hostility against genetically modified organisms is an even sharper illustration of ill-conceived environmentalism. How animals can be modified with the help of cloning techniques was briefly mentioned in chapter 16. Much more important, in terms of economic and political fallout, is the production of genetically modified plants. The technique used to this end is different from cloning and merits a brief digression, because its history shows in exemplary fashion how fundamental research carried out for the sole purpose of understanding a natural phenomenon can lead in totally unforeseeable manner to practical applications of major importance.

A basic discovery opened the way to revolutionary applications

The story begins with a plant disease called crown gall, more commonly known as plant cancer, which is revealed by unsightly outgrowths, or tumors, on affected trees. Early studies of this disease showed it to be caused by bacteria, which were called *Agrobacterium tumefaciens* (tumor-causing) for this reason. This was a major discovery in itself: a cancer-causing microbe! The manner in which the microbe acts proved equally remarkable. It was elucidated in the 1970s by two Belgian investigators, the late Jozef Schell and Marc van Montagu, who found that the bacteria possess a special gene-insertion mechanism by which they inoculate into the plant cell a set of genes that are then incorporated into the cell's genome and expressed to produce the tumor. The investigators further used their findings to develop a procedure in which the bacterial inoculation system is disconnected from the tumor-generating genes and exploited to insert chosen foreign genes into plant cells, in place of its nefarious cargo. Thus, a major basic discovery was turned into a new and particularly powerful biotechnological tool. This example, of which many others could be cited, deserves to be heeded by decision makers who too often are willing to support only investigations aimed at solving a practical problem and likely to lead in the short term to useful and, if possible, profitable applications.

Developed industrially, this gene-insertion technology has led to a large number of valuable applications. Thus the ability to destroy harmful insects, to defend themselves against pathogenic fungi or viruses, or to resist certain herbicides has been conferred to plants as widely different as corn, rice, soybeans, beets, potatoes, and bananas, to mention only a few.

The same gene-insertion technique has also been used to generate nutritionally enhanced plants. Rice, for example, has been treated to produce large quantities of vitamin A, the so-called yellow rice. In another application, the production of allergy-causing proteins has been turned off in soybeans and peanuts.

Other types of changes have been induced for industrial purposes. Potatoes have been engineered to produce large quantities of amylopectin, a form of starch that is widely used to make glossy paper coatings, clothing finishes, and adhesive cement (but renders the potatoes unfit for human consumption). Trees have been modified to make less lignin, the structural component responsible for their hardness, so that they can be exploited more efficiently for the production of biofuels. The list goes on lengthening. Using this technique, any desirable, genetically determined quality can be conferred by insertion of the appropriate gene into the appropriate plant.

One would expect such technology to be hailed enthusiastically, especially as some of its main benefits are expected to favor impoverished and malnourished Third World populations. This has not happened. On the contrary, genetically modified organisms (GMOs) have become targets of particularly vicious attacks.

GMO: an acronym that ignites passions

A salient objection expressed in anti-GMO propaganda has been that a genetically implanted character could be transmitted to neighboring wild varieties and "contaminate" them. The objection, to say the least, is strange. Humans, without knowing the molecular mechanisms they employed, have been genetically manipulating plants and animals for ten thousand years, creating varieties that have little in common with their prehistoric ancestors. A Cro-Magnon individual suddenly transferred into the modern world would be hard put to recognize our corn, wheat, tomatoes, and other cultivated plants, or our pigs, horses, cows, goats, and sheep. All these and other domesticated varieties have been generated empirically by means of hybridization and crossing techniques selected for the sole purpose of producing organisms that were useful and profitable to humans at the time, with no concern for any environmental drawback. Here, for the first time, a change can be introduced knowingly and responsibly, under carefully controlled conditions. And the procedure is categorically rejected!

Another objection is that GMO food could be unfit for human consumption. The term "Frankenfood"—from Frankenstein, the monster-creating scientist imagined in the famous 1818 novel by Mary Shelley—is the scare word invented to highlight the danger. This argument is even less valid than the preceding one. There is no a priori reason to suspect that the organisms carrying such foreign genes would be more toxic than the natural versions of those organisms. The often evoked risk of allergies is, of course, real, as it is with many natural substances. This and other possible dangers can easily be screened by control procedures. Large populations in the United States have been consuming "Frankenfood" for years without ill effect.

In Europe, the Greens have succeeded in manipulating

public opinion to the point that growing GMOs is prohibited in several countries and, if allowed under restricted conditions, as in France, is exposed to lawless destruction. In most European countries, GMO food is authorized for consumption only on condition that the consumers be clearly warned that they are exposing themselves to "risk" at their own peril. A recent decision allowing a small amount of GMO material to be included in commercial food products without warning caused an explosion of protests. People were shown on television expressing their refusal to eat such disgusting "pig's food." Faced with this kind of indoctrination, defenders of rational objectivity are almost powerless.

Even manipulations designed to favor the opposition's aims are rejected. Thus the "enviropig," which has been developed for the specific purpose of protecting the environment against the harmful effects of excreted phosphorus (see chapter 15), has been fiercely combated. A representative of a major, international environmental defense organization called it a "Frankenpig in disguise." Similarly, those who want GMO food banned because of its hypothetical risk of causing allergies oppose manipulations that are aimed at preventing real allergies. Vitamin A–enriched "yellow rice" is still awaiting permission to be cultivated, some ten years after its creation.

Part of the opposition to GMO technology is political and ideological, fuelled by hostility against the perceived ills of capitalism and globalization. The technology is in the hands of a few multinational companies, which, because of the large investments required, were the only entrepreneurs able to develop it. Understandably, they want a fair return on their investment; and their motivation is not always innocent. Thus, creating a plant variety resistant to a given herbicide is particularly profitable to the manufacturer of that particular herbicide, which is the only one among such substances that can prevent harmful herbs from invading the area occupied by the herbicide-resistant plants without attacking the latter. Commercial companies are often accused of coercing Third World populations, making them dependent users of GMOs. The so-called *terminator* gene, which prevented second-generation seeds from being used to produce new crops and made purchase of new seeds necessary every year, did, indeed, fulfill this purpose; that gene has since been removed from GMOs. The fight goes on.

Are GMOs an assault against the sacredness of nature?

The argument against GMOs that most impresses the general public is that they are "unnatural." Inserting a foreign gene into a living organism is viewed as a crime against nature, an attempt at "playing God," an expression used by the heir to the British throne. Nature, in a revival of the doctrine defended by Jean-Jacques Rousseau, is seen as sacred, inherently good, to be revered and not manipulated.

Such sanctification of nature is irrational and rests on no objective argument. Nature is neither good nor bad; it is neutral. Natural selection is blind; it has as much solicitude for the AIDS virus as for penicillin-producing molds, for the scorpion as for the poet. What favors survival and proliferation under prevailing conditions is automatically selected, whatever the nature of the advantaged organism. It is precisely one of humankind's privileges to be able to oppose this blind process and to manipulate nature at will. Humans have, from the onset, interfered with natural processes and exploited these processes for their own benefit, and it was, for them, natural that they did so. It is ironic that, at the very moment when it has become possible for us to take such actions responsibly and knowingly, defenders of the environment oppose progress and advocate keeping to the old procedures, on the pretext that they are more respectful of nature.

Environmentalism has a crucial role to play

This conversion of environmentalism into some kind of religion is regrettable. The menaces exerted by "human progress" on the environment are of extreme magnitude and gravity; they demand urgent measures by all available means. The ecological awakening of the last fifty years has been extraordinarily beneficial in this respect, and the movement it has generated could be of immense importance for the future of the planet. Its political influence should be much greater than it is at present and should carry the adhesion and support of all the populations of the world, all equally concerned by the fate of the planet. Green parties should be dominant worldwide. But for that to happen, their leaders should get rid of their extremist, irrational, and demagogic fringe and work together with scientists, instead of combating them. If we want effective ecological interventions, it is important that these be conducted in rational fashion with the help and advice of knowledgeable and trustworthy experts—which is what most scientists are, contrary to the accusation of venality that is too frequently thrown at them. Such interventions could contribute powerfully to alleviation of human-inflicted damage to our planet and help save it from the menaces that threaten its future. The misappropriation of ecology by mostly well intentioned but poorly informed and inadequately trained advocates is a deplorable perversion of an otherwise hugely constructive and indispensable movement, essential to human well-being on planet Earth.

19
Option 6: Give Women a Chance

Before we close this excursion into the future, a question that has attracted increasing interest in recent years must be addressed: the role of women in the running of human affairs. To keep within the limits of our subject matter, does science have anything to say concerning this key societal problem? Apparently yes, as the evidence suggests that several unfavorable human traits singled out by natural selection are largely associated with maleness.

Combativeness is primarily a male character

Most of the time, wars have been waged by men. Almost invariably, the dominators and conquerors, the soldiers and other fighters, have been men. Women have followed, to provide food, to care for the wounded, and to entertain the warriors with sex. Only exceptionally have they borne arms. This difference in behavior between the male and female members of our species could be written in the genes.

Throughout the animal world, males fight each other, sometimes to the death, sometimes ritually, for the most desirable females, while the females protect their young, or watch, fascinated, this joust of which they are destined to be the prize. The winners of these battles earn the right to entrust their genes to the best females and to give rise to similarly endowed progeny. Thus, the drive to win in battle is, in these species, genetically a male imperative, strongly favored by natural selection.

In contrast, females, especially mammals, are programmed to take care of the young. Their fighting instinct is aroused mostly by menace to their offspring. This fostering function gives females a prominent role in the epigenetic wiring of young brains. Babies are overwhelmingly nursed and nurtured by their mothers—or by grandmothers, nurses, nannies, or other surrogates, almost invariably of the female sex. Women are thus the main purveyors of the stimuli that shape the wiring of the babies' brains. So, they are particularly well placed to change the world for the better.

This is of course an oversimplified way of looking at the sexual divide. Women have their share of competitive impulse. The female sex has given us not only Florence Nightingale, Mother Theresa, and Sister Emmanuelle but also the suffragettes, Margaret Thatcher, and the Williams sisters, let alone the Great Catherine. During the French Revolution, women sat for hours knitting—the famous *tricoteuses*—waiting to watch the executions. According to the popular adage *Homo homini lupus, mulier mulieri lupior,* man is a wolf to man, woman is "wolfer" to woman. On the whole, however, it is fair to say that fighting has overwhelmingly been a man's business.

In most civilizations, women are treated as inferior to men

Favored by natural selection, the subjugation of women by men has been confirmed by custom in many societies, most prominently those that follow the biblical tradition. Whether in Orthodox Judaism, the Catholic Church, or Islam, women are treated as inferior to men, even though this discrimination may be disguised as "respect." The fights around the ordination of women are typical of this attitude. So is the amalgamation by several religions of sex, sin, and feminity, as is the widespread cult of virginity. This atavistic misogyny even goes back to Genesis. Not Adam, but Eve ate the forbidden fruit first. In this pervasive myth, a woman is the origin of our genetic curse.

In line with this cultural tradition, men greatly outnumber women in all leading professions, whether in business, politics, the sciences, or the arts. This is not an inevitable fate. Matriarchal societies have existed. According to certain archaeologists, the Golden Age prevailed in Crete, in the third millennium before our era, under the influence of women of peace. Women have won Nobel prizes or have had works exhibited in museums or performed in concert halls. In politics, especially, they are beginning to emerge, whether by exhibiting typically feminine qualities or by imitating men remains, however, open to question.

The social rise of women in the modern world is an encouraging change

As things stand now, the human world is still largely a man's world. Even the world in which women win remains ruled by competition in every area of endeavor. Whether cooperation and understanding will eventually prevail cannot be foretold. Whether women would run the world better than men is still debatable. But they deserve a chance to try. The problem is: How can women gain power without behaving like men?

Although this problem is far from solved, the present trend is encouraging. There is no doubt that women have, in a mere fifty years, acquired considerably more influence in several parts of the world, as evidenced, for example, by the increase in the number of women professionals or by the shifting distribution of parental responsibilities and of household chores between the two members of a partnership. What is particularly encouraging about this change is that it is taking place largely with the cooperation of men.

20
Option 7: Control Population

In the last analysis, it all boils down to a *population* problem. Most of the ills covered in chapter 12 flow, directly or indirectly, from the fact that there are too many of us now on Earth, and soon there may be way too many. The unbridled multiplication of human beings allows our genetic heritage increasingly to produce its most damaging effects. Initiated many millennia ago, this trend has burgeoned with time, but without reaching tragic proportions as long as there remained on our planet virgin territories to occupy and fresh resources to exploit. The exponential pace of demographic expansion, linked with the extraordinary power of the human species to survive under adverse conditions thanks to its intellectual faculties, was bound to lead one day to a global crisis. Malthus predicted it two centuries ago. Today, it is happening.

The crisis foreseen by Malthus has struck

If there is one action that humanity must urgently undertake to counter its now destructive genetic propensities, limiting its population is truly it. The alarm was sounded in the beginning of the previous century by such insightful precursors as the American Margaret Sanger (1883–1966), a pioneer of birth control, and, later, by her countryman Paul Ehrlich, whose book *The Population Bomb* created a sensation when it was published in 1968. Moves initiated in various parts of the world as a result of these warnings and advances in contraceptives, including the famous "pill," first developed in 1951, have unfortunately had little effect so far. The number of humans on Earth has continued its rise (see fig. 12.1). And in Rome, Pope Benedict XVI solemnly reiterated, in October 2008, his unyielding condemnation of any deliberate method of limiting birth other than abstaining from sexual intercourse during the fertile period of the menstrual cycle. As recently as March 16, 2009, on the occasion of a trip to Africa, he again stressed his opposition to the use of condoms, even to prevent the spread of AIDS.

This, in my view, is more than regrettable; it is irresponsible. We need and should expect that the spiritual head of more than one billion human beings will take the initiative in such a critical circumstance, pronouncing it morally justified, if not commendable or, even, obligatory, to oppose the population increase by all reasonable means compatible with human health and dignity. The issue, of course, rests on what is to be considered "reasonable" and "compatible with human health and dignity," in relation to the gravity of the crisis humanity is facing.

Culling is not a tolerable solution to the population problem

The most drastic means of containing a rapidly rising population rate is the one that hunters sometimes adopt, the culling of herds, preferentially sacrificing aged or sick animals in the process. Applied not so long ago in the Nazi camps, this horrible means is obviously proscribed for humans in all civilized countries. But it is replaced to some extent by the wars and genocides that continue raging in various parts of the planet. With the manufacturing of increasingly powerful weapons of mass destruction, the situation grows steadily worse. There were tens of millions of victims in the last two world wars. One shudders at the idea of what the future has in store for us. But such conflicts are hardly a reasonable solution to our population problem. They are, rather, with starvation and epidemics, part of the nightmare natural selection has in store for us if we do not act against it.

One way or another, the birthrate must be reduced

One means of reducing the global birthrate is by imposing *sterilization*. Certain eugenists in the not-too-distant past have advocated this strategy, not so much for birth control purposes, but to "purify the race" by rendering the "unfit" unable to procreate. This policy is necessarily totally unacceptable. But such a ban does not extend to *voluntary* sterilization, which is feasible today by simple surgical procedures, such as Fallopian tube ligation for women and vasectomy for men. I lack statistics on the subject, but it is my impression that few young people choose to undergo these procedures. Societies may, however, be driven to encourage this choice. Widespread voluntary sterilization could become a particularly simple and effective way to control population growth. This is all the more true because such operations have few, if any, harmful effects on hormonal balance and the procedures are often reversible, so that the ability to procreate can be restored in the case of a couple desiring a child, after the loss of one, for example.

Abstaining from sexual intercourse is obviously a way to avoid procreating. It is the solution of choice for all those who make celibacy a condition of priesthood or monastic life. It is, however, known, through all the recent pedophilia scandals, that the choice of celibacy is not always free of destructive complications and may lead to perverse and seriously damaging behaviors.

Homosexual behavior is another form of sexual activity that generates no offspring, but it is, for the most part, likely to be an inborn, possibly environmentally favored proclivity, rather than a deliberately chosen way of life. A society more tolerant toward this sexual orientation, as obtains today in many countries, could, however, produce a certain positive effect, because homosexuality is perhaps more common than we would be tempted to believe. It is up to us to provide this proclivity with the opportunity to manifest itself without prejudice to the parties involved. But this is an ethical issue, quite separate from our population problem.

The most efficient procedures for reducing the number of human beings remain *contraception* and, as early as possible, *interruption of pregnancy,* including its preventive form, the "*morning-after pill*." It is through such procedures that humanity can best prevent demographic expansion. They are authorized more or less liberally in many countries. But this is not enough. The procedures should be more than just tolerated; they should be *encouraged.*

Limiting births needs to be encouraged

Given the urgency of the problem, political authorities should, with the support of as many moral authorities as possible, take active positions in favor of limiting births. An average of little more than two children per couple would ensure that a population will not increase. Condoms, diaphragms, intrauterine devices, pills, and other contraceptive means could be provided free to all citizens of procreating age, together with the necessary medical assistance for pregnancy interruption (under some conditions to be defined). Everyone's freedom to have children should be preserved, but at a price—perhaps through taxes—that would take into account the impact on society of a population expanding beyond a level that is sustainable within reasonable economic and social conditions.

These can be seen as shocking proposals in a world that has always put children at the center of its preoccupations. The desire to have children is one of the strongest urges written in our genes by natural selection and legitimized by custom. Opposing it goes against our innermost nature. But we must recognize the reality of figures and facts. To do nothing is to yield to natural selection, with the destructive consequences it entails.

The measures I recommend risk being considered simplistic; certainly, they involve all kinds of political, social, legal, economic, and other difficulties. My motivation—what I consider my responsibility as a scientist— is to expose the facts, as I see them, and to try to draw logical conclusions as best I can.

"Anti-alarmists" often point out that the problem is mainly economic, and many observations show that there exists an inverse relation between economic development and demographic expansion. Let the former rise, and the latter will fall. This may be true, but only to the extent that economic improvement goes together with the increased practice of birth control. Without this control, the rising economic level cannot cause birthrates to fall. One would rather expect the opposite because of concomitant improvements in health care.

Recommended policies may sometimes have unforeseen consequences. China is an example. The "one child per couple" policy imposed by the Chinese government has resulted in a worrisome decrease in the number of girls relative to the number of boys, a result of prenatal sex determination, allowing the preferential abortion of girls. This outcome has less to do with the policy itself and more with a widespread social prejudice in China, one that encourages families to produce a male heir. This problem is for Chinese society to address and should in no way serve as a pretext to oppose birth control. Note that cynics could point out that a decrease in the number of girls, but not of boys, is likely to favor a reduction of the birthrate.

China's example shows how local customs may affect policies. It will be up to each culture to decide what means should be allowed, encouraged, or, sometimes, even, enforced to attain the desired goal. Unless measures are taken to curb the human birthrate on a worldwide scale, the "population bomb" is bound to explode, with predictably disastrous consequences. The message from Malthus matters more now than ever.

Epilogue

Two almost contradictory messages emerge from my analysis of the human circumstance. One says that our downfall, our eventual extinction and that of much of the living world, is inscribed in our genes. The other tells us that we possess the unique power to use reason to escape this fate. Whether "original sin" or "redemption" gains the upper hand is impossible to predict. But, at first glance, the prospects are not encouraging.

There is a major difficulty: we must deal with two sharply different time scales. As human beings, we live within the limits imposed by our own lifespan and that of our relatives. On a personal and family level, or even on the social, economic, and political levels, our unit of time rarely exceeds a decade, often less in politics. On the other hand, the perspectives that should guide our actions are measured in centuries, if not millennia or more. Under those conditions, many of us are tempted to echo the words attributed to the Marquise de Pompadour, the favorite of King Louis XV of France: "Après nous le Déluge."

Another difficulty likely to discourage even the most convinced and motivated among us is the feeling of our own impotence. What each of us can accomplish as an individual may appear as of such little import as to seem futile. This is why collective engagement is so critical, why action, under the aegis of political and religious leaders, will be essential.

The situation, however, is far from hopeless, as is made evident by the movement developing around the issue of global warming and climate change. The world is becoming concerned. Measures are being adopted on a national and, even, international level. More impressive, individuals are beginning to act, each in their own little sphere, to economize energy, reduce carbon dioxide emissions, and avoid waste—in short, to incur a small amount of personal discomfort for the sake of a distant common good. This is only one example. One could cite many others, showing that individuals, even masses, can be mobilized for constructive effort. If the world's leaders could become more actively involved in the fight for the survival of humanity and the rest of the living world—the one depends on the other—the dangers that threaten the future of our planet can be deflected before it is too late.

Index